TALK

The Science of Conversation and the Art of Being Ourselves

Alison Wood Brooks

PENGUIN BOOKS

PENGUIN BOOKS

UK | USA | Canada | Ireland | Australia
India | New Zealand | South Africa

Penguin Books is part of the Penguin Random House group of companies
whose addresses can be found at global.penguinrandomhouse.com

Penguin Random House UK,
One Embassy Gardens, 8 Viaduct Gardens, London SW11 7BW

penguin.co.uk

First published in the United States of America by Crown, an imprint
of the Crown Publishing Group, a division of Penguin Random House LLC 2025
First published in Great Britain by Penguin Life 2025
First published in Penguin Books 2026

001

Copyright © Alison Wood Brooks, 2025

Page 185 illustration © Platon Anton/Shutterstock
Page 22 illustration 'Closeness Lines' by Olivia de Recat is used by permission of the artist

The moral right of the author has been asserted

Penguin Random House values and supports copyright.
Copyright fuels creativity, encourages diverse voices, promotes freedom
of expression and supports a vibrant culture. Thank you for purchasing
an authorized edition of this book and for respecting intellectual property
laws by not reproducing, scanning or distributing any part of it by any
means without permission. You are supporting authors and enabling
Penguin Random House to continue to publish books for everyone.
No part of this book may be used or reproduced in any manner for the
purpose of training artificial intelligence technologies or systems. In accordance
with Article 4(3) of the DSM Directive 2019/790, Penguin Random House
expressly reserves this work from the text and data mining exception.

Printed and bound in Great Britain by Clays Ltd, Elcograf S.p.A.

The authorized representative in the EEA is Penguin Random House Ireland,
Morrison Chambers, 32 Nassau Street, Dublin D02 YH68

A CIP catalogue record for this book is available from the British Library

ISBN: 978–0–241–99682–9

Penguin Random House is committed to a sustainable future
for our business, our readers and our planet. This book is made from
Forest Stewardship Council® certified paper.

Praise for *Talk*

'It's hard to overstate the importance of conversation in life and at work. In this brilliantly crafted new book, author Alison Wood Brooks explains when and why conversations breakdown, drag on, or otherwise disappoint, and when and why they flow effortlessly to build mutual understanding. *Talk* will be a welcome gift for anyone who wants to have better, more effective, and more rewarding conversations with friends, family, colleagues, and even strangers' Amy C. Edmondson, Novartis Professor of Leadership, Harvard Business School, and author of *FT* Business Book of the Year *Right Kind of Wrong*

'Appropriately enough, reading *Talk* is like having a conversation with the world's best conversationalist. Alison Wood Brooks brings to life the science of conversation, in which she is a world expert, with the utmost warmth, empathy, and joy. The perfect combination of stories and science, maxims and examples, this book has raised my game as a conversationalist. I'm recommending it to everyone I know' Angela Duckworth, *New York Times* bestselling author of *Grit: The Power of Passion and Perseverance*

'*Talk* is a rigorously researched, evidence-based book by a leader in the field. On top of this, it's a joy to read – a killer combination' Emily Oster, *New York Times* Bestselling author of *Expecting Better*, *Cribsheet*, *The Family Firm* and *The Unexpected*

'For many, one of the lost arts in modern life is conversation. Alison Wood Brooks' wonderful new book *Talk: The Science of Conversation and the Art of Being Ourselves* has the solution. Read this book, and your life will improve' Arthur C. Brooks, Harvard professor and *New York Times* bestselling author

'In *Talk*, Alison Wood Brooks, one of behavioral science's superstars, teaches us how to have more productive, engaging and humane conversations. Witty, science-driven and filled with compelling insights and stories, *Talk* is a modern classic' Ethan Kross, author of *Chatter*

'Alison Wood Brooks has crafted a thing of beauty; a thorough look at how we communicate, why we communicate, and how to unlock the power of communication with this fantastic book *Talk*. We will all be quietly planning talk topics, asking more questions, bringing joy, and finding empathy towards one another after reading this book. There are too many ah-ha moments to count in this book so just do yourself a favor and start reading it!' Anthony Veneziale, Grammy-nominated & Tony Award-winning creator of *Freestyle Love Supreme*

'*Talk* is, by far, the most useful book I've read in a long time. Just two chapters into it, I found myself having more enjoyable and meaningful conversations with everyone from receptionists to the person I live with. The writing is witty and wise, and each chapter is crammed with fascinating stories and astonishing research discoveries. Get this book!' Teresa Amabile, Harvard Business School professor emerita and co-author of *Retiring: Creating a Life that Works for You*

'In this world of noise where true listening is a dying art, *Talk* is a beautifully written and humane corrective. Alison Wood Brooks shows that conversation is a delicate dance, an immensely difficult skill that we can always learn to do better – and, what's more, the learning can be a joy' Joe Moran, author of *First You Write a Sentence*

'Builds a convincing case for practicing and better understanding the elements that shape good conversation. Lucid and pragmatic, this unlocks some of the mysteries of human communication' *Publishers Weekly*

'Brooks' accessible, humorous style will interest everyone from already-experts to those who didn't know that conversation analysis was an area of study' *Booklist*

ABOUT THE AUTHOR

Alison Wood Brooks is the O'Brien Associate Professor of Business Administration at Harvard Business School. Her oversubscribed course 'TALK: How to Talk Gooder in Business and Life' helps MBA students hone their conversational skills. Her research has featured in two of the twenty most popular TED Talks of all time as well as in Pixar's film *Inside Out 2*. She was named an American Psychological Society Rising Star in 2017 and one of the 'Best 40 Under 40 MBA Professors' by *Poets & Quants*.

For Sarah

CONTENTS

Introduction . . . . *xi*

1. The Coordination Game *1*
2. T Is for Topics *24*
3. A Is for Asking *56*
4. L Is for Levity *88*
5. K Is for Kindness *118*
5½. A Quick Pause *149*
6. Many Minds *150*
7. Difficult Moments *179*
8. Apologies *219*

Epilogue *251*

Appendix *257*
Acknowledgments *271*
Notes *275*
Index *299*

TALK

Introduction

YEARS AGO, MY THIRD CHILD—a sweet, scrappy girl who followed on the heels of two raucous older brothers—said "Love you" for the first time and smooched the tip of my nose. My heart swelled. Though I am a serious scientist who is not supposed to feel outsized human emotions, in that moment, a spring of hot, fat tears welled up at the bottoms of my eyes. After fifteen months of feeding, rocking, singing, working, smiling, crying, and (mostly) silent screaming, hearing the words *love you* fall out of her sweet mouth was a miracle. This was a moment for baby books, love songs, fairy tales.

It was only after I tucked Charlotte in her crib, closed the door gently, and crept almost down the hall that it hit me: My daughter had said, "Lick you." As in "I'm going to lick you." Then she'd licked the tip of my nose.

Blame-gaming aside (because, clearly, it was her fault), missteps like these are not unique to conversations with children who are just beginning to acquire language. Conversation is humbling for children and adults alike. We trip and stumble through moments of awkwardness, discomfort, tension, and boredom in hopes of discovering learning, joy, progress, and connection.

And not all conversational problems are simple mis-hearings. Some years before the "lick" incident, over a dinner of soup dumplings in a trendy restaurant downtown, I worked up the courage to tell my good friend that I wasn't sure about her boyfriend. I'd been thinking about it for many months, probably over a year. He was okay, I said, but she was spectacular. I was worried about her happiness, her future. She received my feedback graciously. She didn't seem entirely surprised, and she engaged with me about my viewpoint. When her eyes became a little misty, I switched to a different topic. I knew she'd heard me, and I didn't want to push it.

Two days later, a text from my friend popped up on my phone. Did she want to talk more? Was she ready to open up about the struggles in their relationship or her own doubts? Nope. It was a photo of a diamond ring glinting in the sun. I recognized her hand, the same red nail polish she'd worn two nights earlier at our soup dumpling dinner. Her boyfriend had asked her to marry him, and she'd said yes.

Though I thought our conversation had gone smoothly, and I was proud that I'd worked up the courage to share my perspective, I hadn't helped my friend. And I hadn't just mis-heard her—I had misunderstood her. In her pensive quietude, I thought she was welcoming my views, on the verge of opening up about how she was wondering about him, too. In reality, she had already helped pick out the ring, and she didn't know how to tell me how misguided my advice was.

In response to her photo-text, I wrote back swiftly: "Wow! He's a genius. This ring is gorgeous, just like you. I hope you're absolutely over the moon." Privately, I hoped my enthusiasm would help erase the misguided feedback I'd delivered two nights earlier. I hoped she'd forget how I'd raised the wrong conversation topic, at the wrong time, and misread her cues.

· · ·

In every conversation, we make thousands of fleeting micro-decisions about what to say, how to say it, and when. Some of the decisions go well, and others don't. Even normal, well-intentioned, perfectly reasonable choices can create problems—from tiny fissures to aggressive ruptures—in our connection. Sometimes we can feel that something we've said has negatively affected a conversation or a relationship. But sometimes we don't know that something's gone wrong, or why. Sometimes problems pass fleetingly, no big deal. But sometimes our conversational missteps can have big consequences: from unease, confusion, and awkwardness to hostility, depression, and heartbreak.

Even though we do it *all the time,* conversation is surprisingly tricky and high stakes. In fact, it is one of the most complex and uncertain of all human tasks—one of our most cognitively demanding feats. There are two key reasons for this, which we'll group broadly under the headings of "context" and "purposes."

Context. The music video for Ariana Grande's 2020 song "Positions" portrays the singer as the president of the United States. Everything around her keeps changing. She's in the Oval Office on a phone call, her chief of staff whispering in her ear. She's in a boardroom filled with world leaders energetically debating something important. She's at the podium in the briefing room, backed by press secretaries and aides, addressing rows of eager journalists. She's careening across a snowy front lawn, walking small dogs. The rapid set changes in her music video are a lot like the context changes in our conversational lives. Every factor of our environments—who's there, what we're talking about, where it's taking place, when it's happening, why, and how—can change in quick succession. Even small shifts—someone pulls up a chair, turns on some music, brings out a board game, dims the lights, walks outside, or simply changes the subject—are important. And each shift in conversational context asks us to adjust—to read the room and bring our best self—as we shift nimbly (and not so nimbly) from one moment to the next.

Purposes. One of the trickiest aspects of our ever-changing conversational context is that we all have different, often competing priorities. Our purposes—our reasons for engaging in any given conversation—are vastly complex and dynamic. We might want to give a friend space to cry, avoid agreeing to a low wage, learn something new, be a sounding board, understand what's going on in someone's mind, have fun, vent, persuade our partner to do something for us or to do something for another friend or stakeholder, and so on. Sometimes we have many conflicting goals within the same conversation, and sometimes we have goals we're not even aware of. Imagine a conversation with a direct report about choosing artwork for the office, for example, where we think we simply want to get this task taken care of and order aesthetically pleasing art, not admitting to ourselves that our other key goals include actually sounding like we know anything about art and making sure the direct report doesn't buy the hideous piece he likes. Yes, even in seemingly easy conversations, our partners have their own goals that we need to divine and reconcile.

At its essence, conversation is an ongoing act of co-creation in pursuit of all manner of needs and desires, a stream of micro-decisions delicately coordinated between multiple human minds, as the context changes at every turn. We do it frequently, so it feels like we *should* be experts; in reality, we are amateur builders.

But I bring good news from the world of science! We can learn to do conversation better—we can be more astute at assessing and adapting to the context, and we can be more mindful of our purposes and how they shape what we choose to say and do. And here's the best part: learning to converse even a *little* more effectively can make a *big* difference—for the quality of your close personal relationships and friendships; for how you come across in your everyday interactions; for your professional success; for the impact your existence will have on the world.

Will this book teach you the secrets to splendid party repartee? The key strategies to becoming a savvy and empathetic leader? The verbal tactics to charm your date? Yes! Though perhaps not in the way you imagine. Being good at conversation requires more than using specific words or magic phrases. We all have too many far-ranging conversations to begin to imagine that we can follow the same script—plus, we can't script how our partner will respond. Communicating more effectively doesn't mean using all the right words all the time, or applying a finite set of communication tactics, or avoiding fissures, or pretending those fissures are not there. Conversing well means *expecting* problems, noticing them, and working to solve them as best we can—and, knock on wood, having some fun along the way. Only then might we coordinate our micro-decisions a little better, in a relentless pursuit to understand and delight each other—across all the contexts of our social worlds.

. . .

One of the things that holds people back from sharpening their conversational skills is a dearth of feedback. I once knew a guy in college who videotaped his golf swing, watched it over and over, and worked to correct his stroke based on what he saw. Like many athletes (including my high school basketball coach, who took my team to the self-proclaimed "Shot Doctor"), he swore that it helped. Alas, this is rarely an option when it comes to conversation. You usually can't go into a conversation by asking, "Do you mind if I record this so I can learn from my screw-ups?" This is why we all transform into Narcissus during video calls, battling to tear our eyes away from our feedback-starved selves—and why, on the rare occasion when we watch footage of ourselves, it is equal parts mesmerizing and mortifying. Even if our chat partners held a mirror for us to study our own gestures and facial expressions midchat, our analysis wouldn't escape our distorted self-view: most of us think our conversations are worse than they actually are, focusing on the wrong moves and moments to correct.

This may have been the case with my advice over soup dumplings—I may have been too hard on myself about it. The truth is that I don't know how that conversation made my friend feel. More than ten years have passed since then, and—across interstate moves, professional triumphs and losses, pregnancies and miscarriages—we've never discussed the unperceptive advice I proffered that night. Maybe she cared deeply about what I said, but it wasn't enough to change her mind about her boyfriend, and she didn't want to make me feel worse, so she let it go. Or maybe she was so excited about the impending engagement that she was able to easily brush off her silly friend right then and there. Maybe she didn't hear me at all, her mind wandering happily from soup dumplings to diamond rings while trendy music thumped overhead.

It's hard to know. Very few conversation partners provide high-fidelity, real-time constructive feedback, like "You sounded really angry, and I think it made people uncomfortable" or "That abrupt topic change was hurtful" or "I don't think you should profess your dislike for my boyfriend now because we're getting engaged imminently." They also rarely give positive feedback like "Wow, your lemur joke really broke the ice" or "Your smile is infectious." And they can't give feedback on all the many things we could have said but didn't. It's hard to improve at conversation when we just don't know what we're doing right or wrong . . . or when . . . or why.

Thankfully, and rather uniquely, for much of my life, I was lucky to have precisely this kind of feedback. One time I remember watching myself have a fireside chat with two industry experts on environmental sustainability and impact investing. I was wearing black pants and a nubby tweed jacket, curly hair swept into an updo. I sounded knowledgeable, enthusiastic, and at times, charming. I seemed slightly annoyed when the conversation was dominated by the two men, but I spoke my views assertively, undaunted by the large audience watching and listening. I admired the moments when I smiled and supported the other chatters, and I wondered what I was thinking when, at one point, I cast my eyes downward to laugh. I wasn't sure what I was thinking because it wasn't really me at all. It was my identical twin sister, Sarah.

From the moment I opened my newborn eyes, I watched an uncanny avatar of myself make all sorts of micro-decisions from close range. Growing up, my sister chose many of the same things I would have chosen in the same circumstances—eat the apple, not the banana; dive into advanced math, not history; wear baby oil rather than sunblock (major regret). Our decision-making was so similar that we got the same score on every single exam we ever took in high school. (I know. Creepy.) Of course, we made divergent choices too. She played the flute, and I the oboe. She wore her hair curly; I woke early to press mine straight. She liked three-pointers; I preferred midrange jump shots. Our whole childhood was a bizarre natural experiment foisted upon us by a higher power (and foisted upon our parents, too, who didn't know they were having twins until the doctor whispered, "Oh, my God, there's another one," moments after I was born).

What this meant for communication was that I had not only a built-in conversation partner but also a real-life mirror. I was lavishly furnished with information as I watched an identical copy of myself navigate the social world every day, sitting at the same table in the cafeteria, sprinting and stumbling on the same soccer fields and basketball courts, raising our hands to answer the same questions in chemistry, attending all the same card games and dance parties. I cringed when Sarah made ill-timed jokes or sniped at someone in frustration. I beamed when she fielded difficult questions with blazing competence or launched the whole lunch table into a laughing fit.

I tried to avoid her lapses and replicate her successes, and like many siblings (twins or otherwise), we also gave each other relentless direct feedback. (Fun, right?) Comments like "That was mean" and "Eww!" and "Don't do that," accompanied by approving and disapproving glances, were (and still are) commonplace—habits well worn across countless shared encounters, from putting on juggling shows for our parents and playing cards with the neighbor boys, to back-diving at lake parties and leading chants to hype up our basketball team. I was privileged to have so much conversational feedback that I often wondered how any of us manage without it.

This fascination with analyzing and improving human interaction led me first to Princeton University as an undergraduate (without my doppelgänger by my side), where I dove headlong into the science of human behavior, and then to the Wharton School at the University of Pennsylvania, where I studied how people's *feelings* influence their behavior. I eventually landed as a professor at the Harvard Business School, where I was to teach a course about negotiation and, presumably, to do research on negotiation, too.

Around that time, though, I came to see that so-called "difficult conversations" like negotiations weren't the only interactions tripping people up. People struggle with seemingly easy conversations, too. And seemingly easy conversations can become acutely difficult at any moment, when we stumble into unfortunate topics or poke unexpected, hurtful barbs into each other with our words. So in my research lab, where I've led a team of doctoral students, research associates, and other faculty colleagues, we've rigorously examined not just negotiations and obviously fraught conversations, but all manner of conversations, at large scale, across far-ranging contexts: speed dating, parole hearings, doctor-patient interactions, negotiations, sales calls, instant messaging, and face-to-face chinwags between strangers, friends, romantic partners, and family members. We've video- and audio-recorded people's interactions, transcribed their words, coded their facial expressions and body language, captured their concurrent thoughts, and linked their conversational choices to concrete results. It's a new science of conversation, and it's everything I'd been looking for—because it's the closest I've seen scientists get to capturing how people really *are*.

You want to understand finance? Study how couples talk about money—or how financial analysts interact with their bosses. You want to learn about law? Look at how lawyers talk to their clients and colleagues. You want a clear picture of the art world? Record art dealers talking to their clients and to artists themselves. You want to understand music? Watch bands write songs together, and listen to their chatter while they rehearse.

Teaching and studying negotiation on the one hand and experiencing thousands of conversations on the other led me to an aha! realization: While strategic and technical skills can help people get ahead in many ways, being a successful person is about relationships. And relationships are about *talking*. Good people, the types of leaders we are trying to train at the business school (and everywhere else), are those who understand, connect with, learn from, and inspire other people. At the same time, interpersonal skills are notoriously difficult to teach. Yes, there are classes on communication and negotiation, power and influence, but I felt that those courses were missing something. They teach about communication, but they rarely focus on actually *doing* it. They teach how to exchange and leverage information, but not how to exchange information while also developing sturdy, nuanced, and rewarding relationships.

So much of what we are trying to accomplish in any conversation is conditioned by the tenor and tone of the underlying relationship, even if it's just being established. In practice, people are less likely to exchange accurate information and are more likely to avoid doing so when they don't like or respect each other or don't feel comfortable and valued. We struggle to interact at all if we're too angry, anxious, or bored, seeking the nearest exit rather than engaging to inform and learn. In one study by organizational scholar Sean Martin, people who were instructed to discuss nonwork topics at the beginning of a work meeting were much more likely to learn useful information—about work!—at the meeting, to use supportive language, and to stay in touch with each other weeks and months later. My own research found that people were much more likely to exit negotiations when they were randomly made to feel anxious, and others' research found that they were much more likely to conceal information when they were made to feel mad. How people talk to each other changes how they feel about each other, which changes how they talk to each other, which changes what they know, which changes how they feel about each other, which . . . You get the point. A never-ending loop of information exchange is embedded in the inescapable landscape of our *relationships*.

Luckily, I had a very sturdy relationship with my soup dumpling friend. We'd been quite close for almost ten years before meeting in that trendy restaurant downtown—ten years filled with joy and love, expressed to each other in tiny moments, one conversation at a time. Though I felt embarrassed that I'd questioned her boyfriend just days before their engagement, I suspect she took it as a signal that our relationship was safe and sturdy—she knew that I was the kind of friend who cared enough about her well-being to say anything at all. Even regrettable moments can be overlooked, repaired, or forgiven when the underlying relationship is worth it.

In 2019, I designed a new course to put my scientific yet relational perspective on conversation into practice. The course is called "TALK: How to Talk Gooder in Business and Life," or just "TALK" for short. It's based around the necessary ingredients of a good conversation to exchange information, but also to use conversation for purposes that aren't about information exchange at all: to feel confident, to have fun, to maintain privacy, and to capture the magic of human-to-human connection. The course has been incredibly rewarding and very popular. In fact, it seems that I've hit a nerve. I've taught it to over a thousand MBA students and executives in just four years, with a wait-list I can't accommodate. I'm now approached for advice from people in a wide array of industries—education, medicine, finance, sports—because it turns out that every organization and every industry wants to figure out how to converse better. When the Boston Celtics invited me to become a consultant for their coaching staff, I thought *Really?* but quickly came to see that conversation is at the core of what they do, too. Coaches, players, staff, managers, owners—all of their work hinges on how well they interact individually and as a system. Conversation is the key to the social world.

This book is a natural extension of my TALK course. It aims to help you feel more confident and competent in your conversations—to feel like you're standing on solid ground, so you can take more chances and expand what's possible in the conversations you have every day. Not everyone will be able to take my course, but this book is for everyone—introverts and extroverts alike, for whatever roles you play in your organizations, groups, and families. I wanted to write it because there are so many of us out in the world who are just like my students—anxious but also excited and ready to come alive through conversation. And good conversation *can* make us feel alive. It can stave off loneliness, one of today's greatest threats. It can satisfy us in ways that few other things do. And blessedly, it's not a limited resource. Quite the opposite. The more people who communicate well, the better off we will all be.

• • •

In this book, I will take you inside the world of conversation, just as I do with my students. We'll start in Chapter 1 with what conversation is: a coordination game. The game is surprisingly tricky, with trapdoors and challenges hidden among a maze of decisions, but thinking about it as a game helps us see how every conversation is co-constructed by multiple players, and how conversation can be *fun*, too. The limits on great conversation aren't just how *we* can improve, but how we can help our partners be better at the same time. It isn't a game we win or lose—it's a game we get to play together.

Then, taking a cue from the philosopher Paul Grice, whose ideas about conversation have guided thinking and scientific progress for decades, I'll walk you through a new framework I've developed called "TALK." The TALK maxims break conversation down into four crucial reminders that will guide our entire approach to make conversation more vibrant, enriching, and effective:

Topics, because great conversationalists choose good topics
and make any topic better;
Asking, because asking questions helps us move between
topics and dive deeper into them;
Levity, to keep our conversations from becoming stale; and
Kindness, because great talkers care for others and show it.

Topics and Asking focus on the structure of conversation and how we can make choices that actively steer us in good directions—moving toward pastures that will give us the best chance of achieving what we want to achieve. While we graze those pastures together, Levity helps us avoid boredom with moments of playfulness and fizz, because good conversation requires mutual attention and engagement. Meanwhile, Kindness explores the power of respect and good listening to bring out the best in each other—to make sure we not only feel heard but *are* heard.

It makes sense to move through the TALK maxims in this order, but in practice they are mutually reinforcing. Asking questions can help you switch topics smoothly. Staying on topic might help you ask a difficult question later that you're dying to ask. Sparking levity can allow you to ask deeper or simply riskier questions. Choosing topics with other people's interests and purposes in mind is an act of conversational kindness. Responsive listening with a call-back always gets a laugh.

And of course there are many challenges to all this, like the fact that conversation is a relentless and uncertain decision environment, and we have little control over how our partners play the game. And so, in the second half of the book, we'll explore the things that test and stretch the maxims: how adding more minds creates a coordination kerfuffle; how moments of trouble can arise at any time for almost any reason; and how dealing with these threatening moments is tricky but *totally manageable.*

Finally, we'll get to apologies. We'll gaze out over the sprawling trajectories of our relationships and lives and see how saying "I'm sorry" can make the difference between harming our relationships and pulling us ever closer—so we might get to enjoy the social world together. In an environment where we're bound to make mistakes big and small, apologies may be the most powerful tool we have to save relationships that deserve saving. Only through conversation can we construct and maintain a shared reality with others, creating a private world with each of our conversation partners—worlds that can grow sturdier, richer, and more rewarding over time, or deteriorate into dust.

* * *

A few months after my daughter, Charlotte, announced her intention to lick me, she learned to differentiate between the words *lick* and *love*. Swoon. Now, four years later, the phrase "I lick you" has become a sporadic joke between the two of us. One night, when I was putting her to bed, I told her she needed to go to sleep, and I needed to work on this book. I even let it slip that I was feeling a little stressed. She leaned in for a hug to comfort me. My heart burst as I leaned in to accept it. As I closed my eyes for a sweet embrace, she licked me squarely on the nose.

CHAPTER 1

The Coordination Game

THINK OF A CONVERSATION you had recently. What are you imagining? Domestic chitchat at home? A grid of faces on Zoom or Microsoft Teams, brainstorming? Grocery store gossip? An oozy first date? That sprawling group text chain. A catch-up with your mom. A smoke break in a greasy alley. Pillow talk. Delightfully disordered chitchat with children. A tense work meeting. Polite banter with the cashier. A sincere heart-to-heart.

All are excellent examples. These days, any back-and-forth exchange of words between two or more people counts as conversation. And not only in the soulless world of behavioral science, in which I work, but for most people, everywhere in the world.* *Merriam-Webster* says conversation is "oral exchange of sentiments, observations, opinions, or ideas," Wikipedia describes conversation as "interactive communication between two or more people," and the *Oxford English Dictionary* defines it as "a talk, especially an informal

---

* In this book, we'll focus primarily on synchronous conversation, in which people respond to each other immediately. I encourage you to think about how the principles we discuss—topic management, question-asking, levity, and kindness—may (or may not) apply to less-synchronous interactions (over text, email, social media, and so on) compared to live, bubbling back-and-forth.

one, between two or more people, in which news and ideas are exchanged." Though our understanding of conversation today is pragmatic, the definition of *conversation* was not always so.

Three centuries ago, *conversation* meant something very different and quite specific (and none of the examples above would have qualified!). It was a high art, defined by elevated exchanges on high-minded topics: opera, poetry, politics, freedom. It took place between particular people—the most cultivated aristocrats and accomplished writers and thinkers of the day. For these luminaries, the art of conversation was itself a fascinating topic of conversation—it was the early days of talking about talking. What defined conversation? And what defined a great one, especially? Which nation practiced it best? This was the Age of Conversation, and almost everyone who was anyone had a view.

The philosophers and socialites who took up these questions—David Hume, Adam Smith, Jonathan Swift, Germaine de Staël, and Johann Wolfgang von Goethe, to name a few—agreed that conversation should be mutually "pleasurable" and "agreeable" for everyone involved, which meant it had no room for "strong opinions." In her essay "The Spirit of Conversation" (1813), the celebrated Madame de Staël, who presided over one of Paris's most glittering salons, compared conversation to music and wrote of the former: "It is a certain manner of acting upon one another, of giving mutual and instantaneous delight, of speaking the moment one thinks . . . of eliciting, at will, the electric sparks which relieve many the excess of their vivacity, and serve to awaken others out of a state of painful apathy."

You might be thinking, shouldn't this emphasis on mutual pleasure go without saying? Who wants to host a gathering that only a few people will enjoy? But this seemingly obvious platitude was much less obvious back in the day. In fact, it was pointed. Conversation then meant *enlightenment,* sparkling and spirited intellectual exchange among people who had the quality of mind and the social graces to generate it. And, crucially, it also meant freedom from the rigid hierarchy and rituals that defined court life in an absolute monarchy. In the decades before and after the French Revolution, dissatisfaction with dynastic rule was rampant. And conversation emerged as a fresh alternative to the old world order. In the salons of Paris, where the men and women who inhabited the so-called "republic of letters" gathered, conversation was not performed for the pleasure of the king, according to rules he dictated. Conversation was practiced for the mutual pleasure of the enlightened, according to rules they devised for themselves. *Ooh la la!*

Kant's Rules

Paris was the center of the salons, but "polite conversation" was a topic of fascination all across Europe. In Germany, the great eighteenth-century philosopher Immanuel Kant, author of the age-defining essay "What Is Enlightenment?" (which is still a staple of college courses around the world), was among those who opined on the benefits of polite conversation and its best practices.

For most of his long life, Kant eked out a living as a tutor, teacher, and librarian, hoping to land a full-time university post. His small income relegated him to rooming houses—which meant that, unless he was invited to dine at the homes of friends, as he often was, he had to seek dinner companions in public houses. Despite Kant's reputation for extreme austerity and self-discipline,[*] contemporary accounts portray him as a vivacious and charming guest, whose company was sought after by the town's most illustrious hosts. Kant cherished these private occasions, but after years of supping mostly at local restaurants and hotels, he tired of the boisterous and banal exchanges he found at communal tables. When he finally procured an academic position and was able to buy a house of his own, not long before his sixtieth birthday, he decided to start hosting dinners himself. Finally,

---

[*] Kant very rarely left his small hometown of Königsberg, where he kept such a strict daily routine that his neighbors called him "the Königsberg clock."

he could pick his guests—and his conversation. If you're going to stay in one small, remote town your whole life, you'd better make it fabulous. And he did! Kant's invitations were highly coveted. One excited invitee marveled at having been "asked to table by the King in Königsberg."

Aside from the presence of the Great Man himself, part of what made the dinners special were the rules of conversation he insisted on. Some of the rules were common to the women and men of letters who shaped the Age of Conversation—no interrupting, no monologuing, no talking shop. But many others were peculiar to Kant. Guests should be a mix of ages and backgrounds—professors, physicians, clergy. Never fewer than three guests, but never more than nine— usually five to eight people. After arriving promptly at one p.m., they would follow a three-course "topic menu" starting with light fare— the news of the day, weather, and gossip.

During the main meal, they would turn to serious topics like chemistry, meteorology, natural history, and above all politics (such as the evolving French Revolution, a topic that fascinated Kant)—but never politics *before* dinner. They would conclude the meal with "jesting," lighthearted banter that would elicit laughter and thus, according to Kant, "help the stomach in the digestive process by moving the diaphragm and intestines." There was to be plenty of food and wine, but no beer, no music, no games, and no lulls in conversation. And his number-one rule: no know-it-alls.[*] In fact, he often dealt with disagreements (about the French Revolution or otherwise) by asking that his guests avoid the topic altogether.

In Kant's view, his private dinner parties offered a refuge from the unruly hubbub of public eating places, a safe haven where the art of conversation could flourish. I'll bet that for some of us, the rowdy pub

---

[*] What Kant really said was that to have a "tasteful feast" one must "not let *Rechthaberei* arise or persist." My copy of Kant's *Anthropology* translates *Rechthaberei* as "dogmatism," or for our purposes, being a know-it-all.

conversations would have been more fascinating than Kant's dinners, even if they were less "polite" or rarified. Still, while Kant carefully devised polite conversation in his home, the educated Europeans who belonged to the republic of letters were already, by that time, coming to the conclusion that the art of conversation was in decline. Dispatches from America, where the great social and political experiment of democracy was underway, confirmed striking lapses. The new-world elite were violating many of the rules of polite conversation that their old-world counterparts, including Kant, swore by, not least by talking about vulgar topics like money, work, and themselves.*

In London, too, conversation was becoming more unruly. As the city swelled with the bustle of accelerating trade, commerce, new sources of wealth, and new residents of all classes, the unpredictability of everyday social life was a source of constant commentary and speculation. Streets, markets, shops, parks, and pubs: life outside the highly curated worlds of the dinner party or private salon required chats between people who lacked the shared rules, histories, and rituals that had once ranked them and told them how to behave. Friends and strangers constantly crossed one's path in bewildering profusion, observed the Scottish philosopher Adam Smith, and that bewilderment required "continual" observation of "the conduct of others" as well as constant "adjustment" and "compromise." In "a society of strangers," where people from all classes encountered each other more often and in more places than ever before, conversation was no longer a matter of following explicit rules. It was increasingly a matter of people figuring out the implicit rules and adjusting on the fly. How should they address one another, and what should they talk about? For how long?

---

* Though the connoisseurs of conversation across Europe disagreed about some details of polite conversation—who practiced it best and what the rules should be—they all agreed: Americans were the worst. It wasn't just their accents, grammar, or habitual spitting, deficits that the French aristocrat Alexis de Tocqueville and the once-working-class author Charles Dickens both observed after extensive visits. Americans violated a cardinal rule of polite conversation by talking about business on social occasions—and by talking incessantly about themselves.

How much shared information could be assumed or taken for granted? How was one to distinguish a maid from her mistress if they dressed alike, as they now often did? And did it even matter? *Gasp!*

The Coordination Game

The style of conversation that Smith saw emerging all around him was what behavioral scientists nowadays call a *coordination game:* a situation in which multiple players make simultaneous choices—without communicating. The outcome of a coordination game depends on what everyone chooses to do, not just on one player's choice. Some coordination games are considered *noncooperative*, like the well-known Prisoner's Dilemma, where two prisoners are being questioned about a crime in separate rooms. The prisoners are collectively best off if they both stay silent. They would stay in jail, but not for very long. Their worst-case scenario is if they both snitch. In that case, they'd both be locked up behind bars for much longer. Though it seems obvious that it's best to stay quiet, there's a problem: they get to walk away completely scot-free if they snitch while their partner stays silent. So each prisoner is tempted to betray the other. Hence the "dilemma."

In *cooperative* coordination games, both players are best off if they both choose the same option—like appearing at the same place for a date (when both daters prefer different activities), as in the "battle of the sexes" coordination game, or successfully avoiding a head-on collision, as in the game of chicken. But how can two people make the same choice without communicating? This is why coordination games are sometimes referred to as coordination *puzzles* or *problems*.

Coordination puzzles are a familiar and flummoxing feature of daily life. Imagine you're striding headlong toward a stranger on a narrow sidewalk. You both want to give the other the right of way, but you keep veering the same direction. Or imagine your phone call just got dropped. Who will call whom back? You both want to reconnect, but you're each ending up with the other's voicemail. In both cases, the players make their choices independently, based on their best assumptions and guesses about what their partner will do.

And both cases raise a dizzying number of questions: Which course of action do I prefer? Which course of action do I think my partner thinks I prefer? Is she aware, motivated, and capable of anticipating my needs and accommodating them? Is she thinking about this at all? Which option do I think my partner wants to choose? Which is she most likely to choose? Should I defer to her preferences or insist on mine? Who gets to choose? The walkers and phone callers make independent choices (veer left or veer right, call back or wait for the other to call back). And the outcome of all this mutual mind-reading and guessing—pass each other smoothly or collide; leave multiple frustrated voicemails or return seamlessly to the conversation—depends on what everyone has chosen to do.

In the 1950s, the Nobel Prize–winning economist Thomas Schelling was intrigued by people's attempts to coordinate. To his astonishment, when he asked people where they thought the best place to meet a stranger somewhere in New York City would be, without the chance to talk to each other, a shockingly high proportion of them chose to meet at the information booth in Grand Central Terminal at noon. Even without saying a word, they were able to coordinate because Grand Central Terminal is what Schelling called a "focal point," a solution that people choose by default in the absence of communication because it is stands out as salient. Of course, focal points don't always work. Other common guesses for the New York meetup included the top of the Empire State Building, the Statue of Liberty, and the heart of Times Square. Heading there meant missing all the folks heading to Grand Central Terminal (and getting lost in a mob of tourists). In a coordination game, the potentially endless degrees of guessing can create a madcap, sometimes comical, sometimes infuriating amount of uncertainty and second-guessing.

Conversation is the ultimate coordination game, though not one that Schelling or any other game theorists studied as such. As in classic coordination games, we can't communicate about everything during our conversations. Even while we're busy talking, conversation requires a staggering amount of mind-reading and guessing. We have to think about what other people want and guess what they will do, in an effort to see that it goes well for everyone involved.

But conversations are also quite different from the simplistic choices that game theorists have pondered and that Schelling asked of his subjects. The most notable difference is that conversation demands *more*, much more. More decisions. More mind-reading. It's not just a matter of resisting the temptation to snitch one time or arriving at the same place at the same time (which is already no mean feat). Conversation requires a *relentless* stream of coordination decisions. We have to figure out if the other person wants to talk at all (that's a big if). And if so, then there's the question of what to talk about? When a solar eclipse is happening outside, and the city suddenly goes dark, it's pretty safe to assume that the natural celestial event is on their mind too, and thus it's an easy topic. But have they already talked about it with every other person today? Are they sick of it? And once you've briefly marveled at the eclipse, what next? Can you safely assume that the other person is following the Super Bowl—that it is on their radar, a salient focal point for them, much less that they are rooting, as you and everyone you know are rooting, for the home team? And what about the upcoming election? They must be thinking about it too, but in what way? Who might they vote for, and which issues do they care about? Should you broach these questions directly or avoid them altogether? How will these choices make them feel? What do you want out of this conversation? What do they want? Are you sure?

And then there are the conversations with people you know well—family members, besties, lovers, colleagues. Even intimate conversations require a surprising amount of decision-making under uncertainty. Should you ask about their ailing mother, or will that make them upset? It's not obvious, but let's say that you do ask. How did they react? Are they grateful, reassured, and eager to talk about her, or are they sad, bored, irritated, or angry? The answer to this series of questions will determine if you stay on that topic or switch—and in both cases, how exactly you do so. Make a joke? Apologize? Mention the fancy decor or the flat beer? How familiar or formal should you be? And depending on how things are going, when should you leave?* For our purposes, we'll call these seemingly tiny choices *micro-decisions*. They include every choice talkers make—what they say, and how they look and sound while they say it. Each micro-decision addresses its own coordination puzzle, with its own focal points and with varying degrees of uncertainty.

Kant had no time for conversational chaos, so he designed explicit rules to avoid it. His table-talk maxims set specific parameters—number of people, opening with the weather, politics during the main course, jokes at the end—that would increase the likelihood of coordinating and finding that mutual spark. Despite all the rules, he stopped short of scripting the conversation. Unlike chamber music—the genre popular in his day—conversation can't be scripted. Still, once he took his seat at the table with that evening's guests, just as a lead violinist or conductor might cue the chamber orchestra to start playing with a nod, Kant liked to unfold his napkin and announce the beginning of his conversational enterprise: "Now, then, gentlemen!"

Today, we often find ourselves in situations that lack anything like the rules that people understood during the Age of Conversation. In our own age, the sociologist Arlie Hochschild has called conversation "the jazz of human exchange"—a very different genre of music. Our world allows us immense freedom in how we converse and with whom—on a video call, over email, by text, and with people far and near—which only adds to our uncertainty about how to do it well.

---

* Recent research by psychologists Adam Mastroianni and Gus Cooney suggests that almost none of our conversations end when anyone wants them to. Conversational endings are the final micro-decision to coordinate, which is why they so often feel awkward (and dissatisfying).

This uncertainty can be stressful: there are so many ways to get it wrong! But it also means more freedom and opportunities than Kant could have imagined to get it *really* right—to have fun, to be creative, to do good for each other and the world. Like jazz musicians (and jam bands and freestyle rappers and crowd-work comedians), we can learn the predictable rhythms and patterns of conversation and then improvise, together, to fill them in.

We'll learn to approach conversation as trumpeter Wynton Marsalis approached jazz, falling in love with its messy collisions, using them as springboards to coordinate and bond: "Jazz urges you to accept the decisions of others," he said. "Sometimes you lead, sometimes you follow—but you can't give up, no matter what. It is the art of negotiating change with style. The aim of every performance is to make something out of whatever happens—to make something together and be together."

Our Age of Conversation

In order to create good conversation together, as we move through the world today, the questions we need to navigate extend far beyond the social niceties that philosophers like Immanuel Kant and, later, etiquette experts like Emily Post or self-help entrepreneurs like Dale Carnegie once recommended for us. Our questions are about specific people in specific moments—questions about their *minds*. Is that polite person merely humoring us while they look over our shoulder for an exit? Or are they nodding enthusiastically and leaning in to catch our every word? Were they vague when we asked what they did for a living—or eager to share and swap notes? Was that laugh authentic, erupting from their belly, or tepidly polite? When they said "Oh, I love that," did they really mean it?

In theory, you could just ask others to tell you what they want to talk about, or to clarify what they mean, or to describe exactly how they feel. In certain moments, it's possible to be direct: you can say,

"This is making me sad. Can we talk about something else?" or "What's on your mind?" or "Okay, what should we cover next?" But in many other moments, as with classic coordination puzzles, direct communication is neither viable nor ideal. Some of the magic of conversation depends on the feeling of *naturalness*—the feeling that you've effortlessly and serendipitously alighted on one gripping topic after another, that you've intuitively and accurately read each other's minds. But if you make the rules of engagement explicit, or a matter of duty, that magic disappears, and fast.

Like all coordination games in which direct communication is out of the question, conversation requires incredible feats of self-reading, mind-reading, and room-reading—the constant process of observation, adjustment, and compromise that Adam Smith observed with relish. Those feats become even more demanding when our social world expands to include people who are quite different from us—whose habits, interests, preferences, and values may be unfamiliar and uncertain—and whom we encounter in an ever-changing array of contexts: on subways, at water coolers, in waiting rooms, at birthday bashes, in places of worship, on company retreats, and yes, at dinner parties sort of like Kant's.

Doing Things with Words

This is the modern social world to which the philosopher of language J. L. Austin turned his attention when he and his colleagues began to examine so-called "ordinary language." He showed that we cannot possibly understand language if we assume that its main function is to describe the world around us. Rather, we need to understand what speakers are *doing* with their words: Are they requesting, inquiring, begging, issuing a promise, offering an apology, or taking any number of other actions? His point was that people are always doing things with their words.

Think again of the conversation you envisioned at the start of the chapter—the one you had recently. What were you doing with your words? What did you hope to achieve? Perhaps your answer is one thing: "I wanted to have fun," "I'm just in the habit of responding to people," "I needed to vent," "I wanted to support my partner," or "I didn't want to be rude." What about the other people involved? What were they doing with their words?

Some people may protest this question—they swear they didn't have a reason for using the words they used. But we always have at least one purpose. We're always doing *something*. Otherwise, we wouldn't bother talking at all. Same with your talk partner—they too always have at least one thing that they care about, even if it's to uphold the human expectation that they must respond (the fundamental instinct for conversational turn-taking). During a conversation, the purpose—what people are trying to do with their words—is the most fundamental element that we must discern through self-reading and mind-reading. While self-reading requires some awareness of what you are trying to do, mind-reading means deciphering what others are aiming to do.

The processes of self-reading and mind-reading to determine purposes are remarkably difficult, as a huge array of different purposes can come into play in conversation. Your purpose might be to catch up with an old friend about everything that's happened since the last time you talked, make a decision, have fun, learn about their views, rant about your own, keep a secret, and so on. The potential reasons for conversational engagement are vast. I find it helps to visualize them plotted on what we'll call the *conversational compass* (see the figure on page 13).

The conversational compass organizes what we are trying to do in all the many conversations that make up our social worlds. The *relational* axis runs east–west and captures the extent to which we care about serving the collective versus ourselves. High-relational purposes seek to create value for everyone in the conversation (such as when you want to make your partner laugh, help them solve a prob-

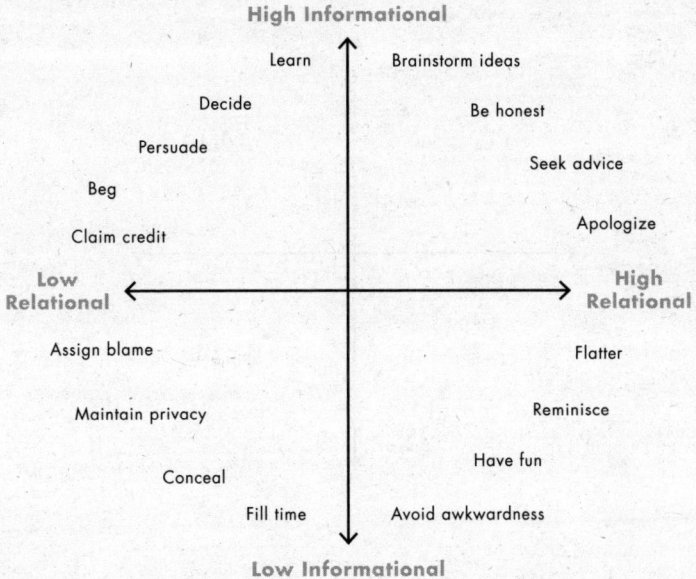

The conversational compass: organizing our conversational purposes

lem, or teach them something new), while low-relational purposes seek to claim value for the self (such as when you want to vent, express your own views, or exit the conversation).

The *informational* axis runs north–south. It captures the extent to which we are aiming for accurate information exchange. Many people assume that information exchange is the main reason we talk to each other—sharing information is why humans learned to communicate, after all. But assuming or over-focusing on information exchange can be misguided. Think of how often you have wanted to guard information rather than share it, how often you have sought to avoid making a hard decision, or how often you have wanted a conversation to feel easy rather than informative. Those are low-informational purposes.

Each of the four quadrants of the compass contains appropriate, worthy, virtuous motives for different moments, which are reflected

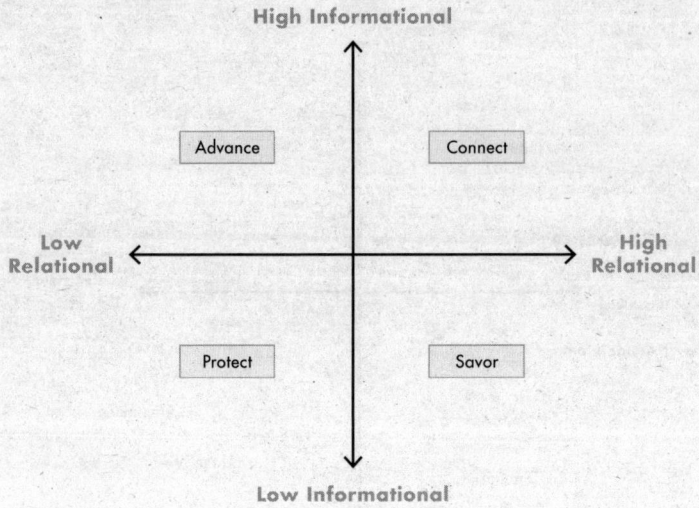

in the positive labels we give the quadrants: Connect, Savor, Protect, and Advance. We live—we do things with words—in all four quadrants.

The universe of possible purposes is unfathomably vast, and there's also no limit to the number of purposes per conversation. Think of a chat with a friend: you might want to convey two easy ideas and one complicated one, hear about the wedding he attended recently, convince him to babysit your kid next weekend, help him decide what to wear on his upcoming date, avoid hurting his feelings, come across as warm and competent, and laugh together as much as possible. Also, you must end the conversation for a scheduled work call in fifteen minutes. The figure on page 15 shows where I might place those motives on the compass (although ultimately, placement is up to the user).

The farther apart your motives are on your compass, the more tension you'll feel between them: it's hard to broach the subject of babysitting when you want to hear every last detail about the wedding; it might be hard to give valuable fashion advice without hurting your friend's feelings; it's hard to cover everything and laugh as much as

THE COORDINATION GAME

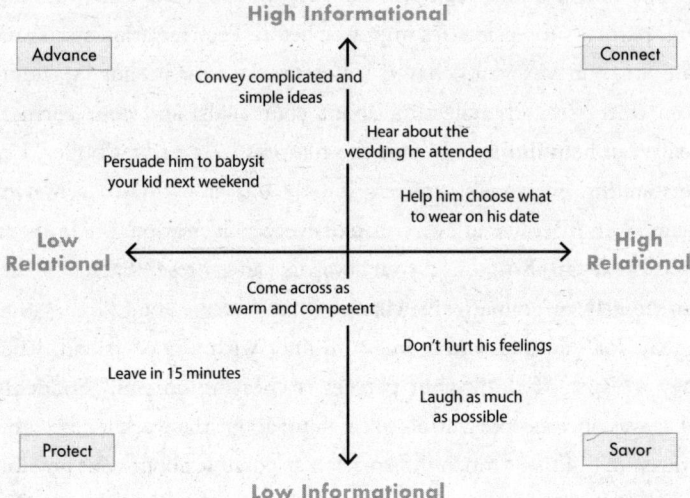

possible in fifteen minutes; and so on. Which of these goals will you prioritize, and how?

Things become trickier still when you realize that your friend has his own conversational compass (everyone has their own!) and that some of his priorities conflict with yours. In Schelling's New York City coordination game, the people involved shared a common aim—to meet up at noon. But coordination games often involve *noncooperative* aims, like the selfish temptation to betray your comrade in the Prisoner's Dilemma, and this is true in conversational coordination, too. Imagine that your friend doesn't want to babysit your kid next weekend, or to talk about the wedding (let alone laugh about it) because the groom didn't show up. And he would love to keep you captive as his personal stylist and loving therapist for the next three hours (at least), rather than abide by your fifteen-minute deadline. Whose desires do you prioritize, and when? How much should you accommodate your friend's needs before you get to satisfy some of your own? How can you figure out which of your purposes are compatible, and which ones aren't?

The conversational compass can help you sort out what your and your partner's top priorities might be before a conversation starts, and to figure out why you behaved a certain way after it ends. Spending even thirty seconds reflecting about your goals, and your partner's goals, can help immensely. But the compass isn't a silver bullet. Understanding everyone's interests is only one step toward achieving them. Why? Because at every turn of every conversation, the immediate context can change, and your compass can get reset. Each new turn can slightly, or dramatically, change what you care about.

Say, for example, you're out to dinner with a good friend, when they whisper, "I think your partner is cheating on you." Suddenly, what was once a conversation to catch up about the weekend becomes a quest to dig up as much information as possible about your possibly cheating partner. Or imagine you're celebrating with a colleague after a big presentation at work, when they say, "I didn't love the last few slides in your slide deck" or "I wasn't sure I agreed with how you answered Howard's question." Suddenly, a conversation to celebrate an accomplishment becomes an opportunity to seek constructive feedback and learn (or to get very annoyed at your co-worker).

We must engage not only in self-reading (understanding ourselves) and mind-reading (understanding our partner) but also in room-reading (understanding the ever-evolving context around us). The improvisational, constantly changing nature of conversation is what makes it both difficult and exciting. You never know exactly what will happen next—you can only use the resources at hand, what you do know, to puzzle it out.

Grice's Conversational Maxims

How people manage to coordinate in this relentlessly evolving environment was the question taken up by J. L. Austin's contemporary Paul Grice, one of the most influential thinkers you've never heard of.

Grice spent decades developing his thoughts on conversation, then finally presented them at Harvard in 1967. In his lectures—which were published shortly after his death as *Studies in the Way of Words*—he laid out the key elements of his theory of conversation. The central idea was the "cooperative principle." This was a bold new look for what he'd previously called, simply and modestly, "helpfulness." And it has sparked endless debate ever since.

Did Grice really think that conversation was motivated by a spirit of cooperation? What about liars, negotiators, bigots, and dodgers? What about all our competing purposes and noncooperative coordination decisions? But Grice was not so naïve. He was suggesting that, at a minimum, conversation requires an element of cooperation. Even when we lie to someone, we still have to communicate with them. And to communicate, we must cooperate. Just as most of us would hold the door for the person behind us, so too do we adjust our conversational contributions to make it easy for our partners to follow them. We take turns speaking and not speaking. We accommodate each other in ways small and large to have successful conversation—even when we hold opposing positions or purposes.

Supporting this cooperative principle, Grice named the assumptions he had once casually described as "what any decent chap should do" as *maxims*. But unlike Kant's rules, which were designed to regulate conversation by providing explicit guidelines, Grice's maxims were intended to capture the implicit rules that tacitly guide people in practice. He gave each maxim an imposing name. The Maxim of Quality: be truthful. The Maxim of Quantity: be concise. The Maxim of Relation: be relevant. The Maxim of Manner: be clear. If these maxims were pursued vigorously, Grice believed, they would result in a conversation that exchanged as much information as needed and no more; that conserved everyone's time, energy, and attention; that didn't veer off topic unnecessarily; and that eliminated obscurity, ambiguity, and misunderstandings.

Alas, in the practice of conversation in our real lives, we violate Grice's elegant maxims constantly. We aren't—and can't be—perfect coordinators.* You don't need to be an expert on conversation to realize that Grice's maxims don't account for the messiness and irrationality of real talk. We aren't always *truthful*. Sometimes we prioritize kindness over honesty because our partners want affirmation and empowerment, and we want to give it ("You look great in that!"). We can't always be *concise*—in some cases our partners need us to fill awkward gaps (as Kant did for his guests), and in other cases brief answers might arouse suspicion or seem rude. We shouldn't always be *relevant*—there are so many wonderful irrelevant topics to discuss: "Have you ever tried a white pinot noir or a bubbly red like Lambrusco?" "Your fly is down!" "Can I tell you about our ice skating excursion?" "I love your stubble!" Good conversation invariably involves a mix of relevance (staying on topic) and irrelevance (jump-cutting to brand-new topics). And we certainly can't always be *clear*. No amount of effort can guarantee that you won't misarticulate your ideas, or that you could ever eradicate *um* and *uh* from your vocabulary. It turns out that those filler words are important—they warn our talk partners about our own uncertainty. We're all making it up as we go along.

A Brief History of Eavesdropping

Around the same time that Grice was laying out his theory of conversation, other scholars were beginning to record actual conversations "in the wild." The renowned sociologist Erving Goffman was one of several social scientists who, in the mid-twentieth century, tried to observe how real people really interact.

---

* Don't worry. Grice was fully aware that his maxims were unrealistic. In fact, part of why he devised them was to highlight the difference between unintentionally falling short of the maxims versus intentionally "flouting" them. He knew that when we intentionally break the rules—subtle, alluring, clever, unique, indirect cues that our partners also understand—conversation is often at its best.

Goffman was a gifted eavesdropper.* By installing himself in corner stores, asylums, casinos, and people's homes, he uncovered the rituals that seemed to shape our everyday interactions and our endless efforts to save face (by which he meant avoid embarrassment and humiliation). Scores of linguists and sociologists took up Goffman's fascination with the small-scale interactions of everyday life. Eventually, this generation of sociologists and linguists founded what became known as Conversation Analysis. They tape-recorded conversations using a single giant microphone, then typed up the transcripts by hand, often straining to hear through background noise. Poring over the transcripts, they carefully dissected conversational habits like turn-taking, airtime-sharing, overlapping talk—and even, in one recent study, sniffing—and devised elaborate methods of notation.

Their methods began with a focus on casual conversations, but over time they studied interactions in doctors' offices, courts, law enforcement offices, helplines, and classrooms. They analyzed one conversation at a time, or a small collection of cases. Their meticulous analyses were crucial in highlighting how turn-taking is the fundamental structure of interaction, how people manage the impressions they make on others, and how they subtly signal misunderstandings and sometimes work to repair them. But still, these scholars couldn't say if the conversations they analyzed were going well or poorly, really. To make claims about whether a conversation is going well, you need to know the participants' purposes—what they are trying to do with their words. And to make bigger claims, you need to study a lot of conversations. Bigger claims required bigger data.

---

* For his graduate research, Goffman took up residence at the remote Shetland Islands. Later, he passed himself off as a pit boss in Vegas, all in an effort to listen in. However, little is known about Goffman's personal conversational style. Famously averse to personal disclosure, he forbade his lectures to be tape-recorded, did not allow his photo to be taken, and gave only two known interviews on the record. He disavowed research in which scholars turn their attention to themselves—"Only a schmuck studies his own life."

And so began the Digital Revolution, offering new technology to record conversations and transcribe them in mere minutes. Now, in the early twenty-first century, automatic recording and transcription tools have alleviated much of the logistical burden of recording face-to-face conversations that once constrained Goffman and his successors. Even more remarkably, the advent of machine learning and another technology known as natural language processing (NLP) means that researchers can now analyze all their data differently, too. These developments, simply put, have been transformative. Spoken and typed words, traditionally viewed as fuzzy, soft, or merely qualitative, can now be quantified—NLP helps us treat words like numbers.

What's so great about these scientific developments? Well, behavioral science now offers better insight than ever into who we are, how we really work, and how we might converse more effectively. Take, for example, gender. Rather than theorizing about gender differences in communication or studying a couple men and women chatting, we can learn—by observing thousands of people—that men and women are equally talkative (psychologist Matthias Mehl recently found that both sexes speak about sixteen thousand words per day on average) and that men and women can communicate in different ways and in different places. For example, the psychologist Gillian Sandstrom and her fellow researchers recently found that women ask fewer questions than men in academic seminars, especially when men have asked questions first, but in my own research on heterosexual dates, I have found that women tend to ask more questions than men. Based on research like this, we can figure out why people talk in certain ways, and we can measure how their conversational micro-decisions influence longer-term outcomes. Did the seminar attendees talk to each other afterward? Did they collaborate? Did they uncover groundbreaking ideas? What are their salaries? Did the daters see each other again? Did they marry? Is anyone happy?

This era of discovery is very new, and there's *a lot* yet to explore. But researchers have started to unlock the secrets of the social world, and we have arrived at some concrete answers. Those concrete answers have revealed that people need to play the coordination game in *all* quadrants of the conversational compass while embracing conversation's uncertainty and complexity.

And real conversation *is* uncertain and complex. The transcripts my fellow researchers and I study are *messy*. Unlike most of the conversations we see on sitcoms and in movies, real conversations don't have tidy scripts. In fact, transcripts from natural conversations usually seem nonsensical, with truncated, half-finished ideas and circular, incomplete logic that's punctuated variously by expressions of love and defensive jabs. And—though Kant would be rolling in his grave to hear it—that's okay! It's how we really are with each other. Conversation—the task humans have been learning to do for over two million years, and that we've all been learning to do since we were toddlers—is trickier than it appears. We can still benefit from explicit rules of engagement (like Kant's) and from understanding what rational talkers might do (Grice), but now, more than ever, we need simple, science-backed help to take the skills we already have and point them in the right direction.

The TALK Maxims

This is where TALK comes in. The TALK maxims—Topics, Asking, Levity, Kindness—are a set of reminders to help us navigate deftly toward our goals across all four quadrants of the compass, especially when many of us tend to drift in the wrong direction and miscoordinate. I deduced these maxims based on ten years of studying conversation, but each maxim is also *intuitive:* when you think of a conversation that went badly, you'll find glitches in some or even all of the maxims.

My work examines whether people can learn successful conversational behaviors and execute them in practice. And I have happy news: They can! I know because I've seen it happen over and over, not only in my research but also in my teaching. In my course at Harvard, the students practice talking, not in contrived roles like "buyer of factory" or "owner of home," but as themselves. We choose topics, ask questions, consider our goals, lift others up, and talk about the challenges and opportunities of applying those skills across diverse situations.

By the end of the course, I have witnessed transformations. Extremely serious students become a little more jovial. Quiet students gripped by the fear of speaking begin to contribute to group conversations. Already-confident talkers increase their precision. And they all do it with more patience and acceptance, and less judgment, of others' shortcomings, and they show more affirmation and acknowledgment of others' triumphs.

Not only have I observed these changes in my students anecdotally, but I also *measure* them. At the beginning and end of each semester, I give my students the same task: have a ten-minute conversation with another student about a difficult experience in their life. At the end of the semester, we analyze the transcripts of these conversations and compare them, so the students can see the shifts in their behavior. Woven into their words, we find more confidence, engagement, and connection. And praise be, these young people come to simply feel better about the sometimes-dreadful, sometimes-delightful prospect of conversation.

Are you ready to join them? Enough throat-clearing then.

Let's *go*.

THREE KEY TAKEAWAYS FROM CHAPTER 1

The Coordination Game

- The idea of conversation has evolved over time and place. Today it helps to think of conversation as a **coordination game.**
- Conversational goals can be plotted on the **conversational compass** along two dimensions: informational and relational.
- The **TALK maxims—Topics, Asking, Levity, Kindness**—are reminders to help people achieve their goals, one conversation at a time.

CHAPTER 2

T Is for Topics

WE WERE ABOUT TO START, and the room was filling with students. As I opened my slide deck, a guy named Josh, sitting near the podium, flagged my attention.

"Why did you make us brainstorm topics last night?" He sounded more than a little skeptical. I'd asked everyone to write down at least five topics they could discuss with anyone else in the class.

"Was it hard?"

"Nope, quick. But I'm not sure I love the idea of *preparing* for conversation."

The assignment to prepare topics doesn't take long, but it always makes some students uncomfortable. Piquing their discomfort and curiosity is my job, so I raised my eyebrows and gave him a smile that said, *I appreciate your skepticism. Stay tuned.*

It was Chat Circle Day, an exercise I run near the beginning of the semester. I've overseen the Chat Circle many times, and whether on video chat or in person, it's always the same: some people find the idea of *planning* a conversation deeply off-putting. Even people who meticulously design agendas for work meetings, or who diligently rehearse their responses for job interviews, often consider prepping for a casual conversation to be fake, manipulative, embarrassing, unnecessary—and worse, unproductive. Many of them are convinced that prepping will backfire, that it will make their conversational skills seem stilted and wooden—that is, *rehearsed* and likely to alienate the other person.

For my students, the Chat Circle has many purposes—to allow them to get to know each other; to get used to recording conversations they can learn from later; to practice being more conscious of their own and others' topic choices; and to have some fun. It's still only the third day of class—the first major exercise in the whole course—and I leave the goals purposefully ambiguous. So like Josh, they're all curious: What are we up to?

As the rest of the students ambled in, they saw that the chairs had been arranged in two big concentric circles, with the chairs in the inner circle facing the chairs in the outer one. As part of the exercise, the students would have five consecutive conversations, each five minutes long, with five different partners, all of which would be recorded and transcribed automatically on their phones. The main goal, I said, was to *enjoy their time together.* I told them they'd rate the enjoyableness of each conversation after it ended, that they'd listen back to their chats after class, and that next time we would review what made some conversations more enjoyable than others.

The first round of conversation was a warm-up. They didn't even have to choose topics—I did it for them. Every minute I announced a new topic, such as:

"If you could teleport anywhere, where would you go?"
"What is one thing you're good at but don't like doing?"
"What movie or TV character do you think you'd be fast friends with?"

When the five minutes were up, they opened their phones to take a two-minute survey. *To what extent do you agree with the following statements about your chat?* the survey asked. They indicated their responses on 1 to 7 scales: *It was enjoyable. It was awkward. We laughed a lot. I controlled the topics. I liked the topics. I raised one of the topics I prepared last night. I felt confident. My partner was attentive.*

After this first round, the students in the outside circle shifted so that each one assumed a new position and a new partner. For the next four rounds, the students were in charge of choosing topics with each other. Before each round, I told them they could rely on their prepared list of topics if they wanted to, but because I knew many of them were uneasy about topic prepping, I also reassured them that they didn't *have* to raise them. They could just go with the flow.

After a few rounds, just about everyone seemed to be into the exercise—even skeptical Josh. By the end, after the final round, the students were exhausted, their chairs, all on casters, drifting akimbo. The Chat Circle had become a Chat Carnival.

"You did it!" I was truly proud of them. Maintaining sustained and focused conversation for twenty-five minutes can be hard work. They applauded themselves, too. Even without food or drink, they'd had a good time. We rolled our chairs into one big circle to debrief.

The students this year—same as every other year—were surprised by how much fun they'd had. They were grateful for the chance to get to know their classmates a little better.

Yet, as happens every year, some students continued to push back on the idea of prepping for conversation, even after the exercise was over. A few admitted they hadn't raised any of the topics they'd brainstormed the night before because they'd simply forgotten them. Others, like Josh, didn't raise them for other reasons: maybe they didn't want to or didn't feel like they needed to.

I assured them, "No problem. You didn't have to!"

But then it was my turn to push back. "When I walked around the room during the exercise, the students with their topic lists out didn't *seem* stiff or impersonal. They were laughing. They were engaged. They seemed to be clicking. What's going on here?"

The floodgates opened. Many students said they were surprised by how much they'd gained from preparing topics. "I didn't really think I'd bring up my prepared topics, but I'm glad I made that list in advance," said a redhead with an I VOTED badge pinned to her lapel.

"I talked about the same prepped topic in every conversation, and it was fun every time," added another, in an HBS zip-up jacket.

Another student, in chunky glasses, gave one of my favorite responses: "I didn't really think much about my topic list, but I ended up using it a lot."

"What do you think?" I asked Josh privately as he packed up his notes at the end of class.

He told me he still wasn't 100 percent convinced about the idea of topic forethought, but he had enjoyed an "amazing" conversation with a classmate about Dr. Manhattan from the TV show *Watchmen*. That classmate happened to be walking past us just then, so I asked her about it. To Josh's surprise, she said it was a topic she'd prepped and pulled from her list.

In the next class, I shared the students' data—the average ratings of their conversations, including the variables that correlated highly with enjoyment. The results were the same as with every other Chat Circle: people rated their conversations with prepped topics as far more enjoyable than conversations that were completely spontaneous.

For the rest of the semester, all the students continued to practice topic prep inside and outside class—first as required assignments, then later by choice. Through repeated practice, even Josh saw that a conversation with a list of possible topics in his back pocket was a remarkably better experience than one without it. By the end of the course, over 90 percent of the class rated topic prep as one of the course's top-three most important takeaways.

But why, in practice, would topic prep not produce the stiltedness and sense of alienation that Josh worried it would? Could thirty seconds of forethought really make such a difference? To answer these questions, we must start by appreciating how surprisingly tricky it is to manage topics in general.

The Topic of Topics

In his powerful 1965 memoir, *The Reawakening,* the Italian writer Primo Levi recounted his arduous return to Italy after being liberated from the Nazi death camps. He described meeting Mordo Nahum, "the Greek," a fellow Jewish man, who beguiled the Italian soldiers (his former enemy combatants). Levi was amazed by the man's talents: "Mine was no ordinary Greek, he was visibly a master, an authority, a super-Greek. In a few minutes of conversation, he had accomplished a miracle, had created an atmosphere." Not only did Nahum "possess the right equipment; he could speak Italian," but "(what matters more, and what is missing in many Italians themselves) he knew *what to speak* in Italian. He amazed me: he showed himself an expert about girls and spaghetti, Juventus and lyrical music, the war and blennorrhea, wine and the black market, motor-bikes and spivs." Across far-ranging conversational ground, Nahum made fast friends of his sworn enemies.

Of course, we're not all so deft. In 1939, just a few years before the Greek was beguiling the Italians, the British writer Virginia Woolf recalled a disappointing lunch with the poet Stephen Spender. His "loose jointed mind" wafted over topics that were so "misty" and "suffusive" that when she tried to write them down in her diary later that day, she could barely remember them. "We plunged and skipped and hopped—from sodomy and women and writing and anonymity and—I forget."* The encounter derailed any potential collaboration between the two.

There are undoubtedly many reasons for their failure to connect—awkward turn-taking or distractions perhaps. But how did the Greek "create an atmosphere" in the unlikeliest of situations, among former enemies, while Woolf and Spender, members of the same country, class, and clubby literary circle, failed to click during an easy lunch?

---

* I like imagining Spender's topic prep with a little notecard that says "sodomy" on it.

Between them, Levi and Woolf underscore a fundamental factor: *topics*. Just as in my students' conversations in the Chat Circle, some topics took off more than others. The Greek's talk of spaghetti, wine, football clubs, and motorcycles landed more memorably than Spender's "misty" talk of writing, anonymity, and "I forget" with Woolf.

The words we speak and type and sign to each other—*verbal* conversational content—are what makes our conversations conversations. Nonverbal content (like facial expressions, body language, and hand gesticulation) and the acoustic properties of our voices (like tone, laughter, speed, and pauses) can matter tremendously, too, but interaction isn't conversation without verbal exchange—without *words*. Though we can parse our words in many ways—into phrases, sentences, paragraphs, turns, syllables, and so on—it's the topics that we as talkers choose, follow, manage, and remember. That makes topics very useful when we are thinking about how to get better at conversing. Topics are what we talk about, but they're also much more. Technically, they are chunks of thematically related conversational turns that convey shared meaning across speakers. Unlike words and phrases, which unfold very rapidly, topics tend to unfold at a slightly slower rate that allows our minds to keep pace.

As we've learned, conversation is a complex coordination game. And in the coordination game that is conversation, topics are the fundamental pieces. We intuitively understand that *what* we talk about determines *how* the conversation goes. Our first TALK maxim is Topics because they are the building blocks of conversation, and mastering them is essential. And mastering topic management, as with all aspects of the coordination game, requires self-reading about what you want to talk about, mind-reading about what your partner wants to talk about, and room-reading about what will work.

The False Choice of Good Topics

Like beauty, what counts as a good topic is in the eye of its beholder(s). Girls and spaghetti went over well with the Greek's new Italian friends, but that topic might not have helped Spender in his quest to engage Woolf. One person may find the topic of vintage cars scintillating while others find it impenetrable, boring, annoying, intimidating, or irrelevant. Ditto for every other topic.

Still, there's notable agreement as to which topics would be good, in theory, to discuss with a stranger—and which ones wouldn't. It's a discovery that my fellow researcher Mike Yeomans and I made when we asked a thousand people to rate a list of fifty topics based on how much they thought they'd enjoy talking about each one with a stranger. The topics ranged from the blandly inoffensive ("Have you read anything interesting recently?") to the intense ("When did you last cry in front of another person? To yourself?"). The survey respondents preferred topics that seemed to offer a clear path forward without stepping on toes or triggering negative emotions—questions like "Do you like where you live, or do you want to move eventually?" "What would constitute your perfect day?" and "What are your favorite [songs, foods, TV shows, movies, books]?" Meanwhile, most of the respondents seemed to share the intuition that topics related to sadness and death are aversive. "Of all the people in your family, whose death would you find most disturbing? Why?" came in at the very bottom of the list.

But here's where it gets complicated. The topics that our respondents thought would theoretically be best for conversation weren't necessarily those that people most enjoyed discussing. When we asked people to actually talk about those topics, we found something surprising and encouraging. Many of the aversive topics, to which people gave the lowest ratings on the survey, turned out to be the ones they relished most during their conversations. In fact, the topic "When did you last cry in front of another person?"—which our respondents ranked second to last in their predictions of enjoyment—turned out to be the one that people often enjoyed discussing. Just because a topic seems sad doesn't mean discussing it will make people feel sad.

It's almost impossible to tell in advance whether a conversation topic will be good or bad. Especially with people we don't know well, it's hard to know until we try. Our preferences for conversation topics are categorically different from our preferences for food or colors or movies or shoes, because topics—and all aspects of conversation—are *co-created.* If you hated talking about breakfast with your uncle yesterday, you might think you just don't like talking about breakfast—that it's a bad topic. On the other hand, talking about breakfast with your college friend or a different uncle or the same uncle at a different time might actually be amazing. Who knew Uncle Fred wooed Aunt Flo by making her an herby omelet with goat cheese in 1971? Who knew your bestie could make you laugh so much about the time he burned French toast in his mother-in-law's beloved pie tin? It's not the topic of breakfast itself that determines how much we enjoy talking about it. Rather, it's ourselves, our partners, and all the particularities of our conversation with them.

I'd Rather Manage

Rather than thinking of topics as good or bad in the abstract, it's helpful to develop topics based on your relationship with your talk partners, their relationship to the topics—and everyone's goals.

My research shows that having good conversation is not about choosing good topics; it's about making any topic good through *topic management.* In every conversation, we constantly search for mutually rewarding topics, together. This relentless process of topic management—choosing and steering topics during conversation—is yet another reason the coordination game is so tricky. At every turn of every conversation—every time someone speaks—they have the power to decide whether to stay on the current topic, switch to a new one, call back to an old one, or try some combination of these options. These topic micro-decisions—how we steer the conversation, and how we choose to accept and reject our partner's bids for topics—determine what treasure we're able to discover during our conversational time together (because, of course, it's not just about choosing topics, but also deciding what to say about them).

Prepped for Success

Topic management is why topic prep helps. Think of it as a cheat code that can help you navigate the coordination conundrum.

Most of us, like my skeptical student Josh, balk at the idea of brainstorming topics ahead of time. Good conversationalists don't *plan*, right? They're just naturally talented. Intuitive. Empathetic. Charismatic. Quick-witted. They're genetically predisposed to create smooth, spontaneous connections while the rest of us struggle. My students often express admiration for the charismatic people they know and attribute their charisma to natural talent: "My friend is so smooth. He always seems to know just what to say. I want to be more like him." This habit of thought—watching others thrive and attributing it to natural talent rather than accumulated effort or hard work—is the Myth of Naturalness: when someone displays their skill, it looks effortless, but we come to believe that it *is* effortless for them and that it should be effortless for us, too.

The Myth of Naturalness fuels many problematic fires, including our resistance to topic prep. My research shows that before a conversation, 27 percent of people report spending at least five minutes (and often much longer) deciding what to wear, while only 18 percent report thinking about what they'll talk about. Some people believe that thinking about topics beforehand will be distracting during the conversation, that they'll be too focused on trying to remember what's on their topic list. They're afraid that if they forget, they'll get flummoxed and distracted and introduce awkward breaks and gaps. More than half (53 percent) of the people I've surveyed think topic forethought is simply unnecessary. They feel confident they'll know what to talk about, especially in casual conversations with people they know well. Just over a third think that topic prep will make a conversation feel forced, like they're working through a checklist or following a strict script (34 percent). And a whopping 50 percent believe that brainstorming topics will *decrease* their conversational enjoyment, while only 12 percent think it will increase their enjoyment.

The vast majority of us, then, count entirely on our ability to choose topics in the moment, during the turn-by-turn chaos of conversation, even as aspects of the context and our goals are shifting at every turn. Psychologists call this kind of fast, instinctive, automatic, and emotional processing "system 1 thinking." It's our default operating system. During conversations, system 1 thinking guides topic management. We don't realize it, but we all make intuitive decisions about topics *every time we speak.* At every turn of every conversation, we are choosing to stay on topic, to subtly drift, or to abruptly jag to a new topic—and deciding what to say on that topic—and our talk partners are doing the same thing.

The tricky part is that our system 1 thinking, while it is powerful, often leads us astray. We instinctively grab what's right in front of us—whatever topic springs to mind. But the topics that come to us spontaneously may not make for the best conversation (far from it). We talk about things that are top of mind—the fact that it's sunny or cold or raining, or that a random guy is standing over there, or that the salty appetizer just arrived—rather than the questions, stories, and ideas we truly care about. We talk about what we know or want to know or what's easy, rather than try to learn what our partner knows or might want to know, or broach trickier subjects, or search for things we're *both* curious about. We try to prove how great we are or how right we are (our natural egocentrism), rather than try to figure out what we can learn from our partner or what will produce the strongest sense of engagement. Guided by the Myth of Naturalness—the belief that we should trust our gut—we risk drifting through conversation *too* intuitively, rather than trying to identify what could work better to achieve our goals.

Proceeding without forethought is tempting because of the illusion that conversation should be easy and effortless. But in truth, conversation is neither easy nor effortless, even for Primo Levi's Super Greek or the charismatic friends my students admire. Good conversation—like good acting, writing, music making, trapeze flying, or any other kind of skilled performance—requires a combination of instincts and deliberative effort, even if the latter is invisible to onlookers. Psychologists call that slower, more deliberative, and more logical effort "system 2 thinking." To become great conversationalists, we need to take advantage of both snap-judgment system 1 and deliberative system 2. We need to speak intuitively in the moment but also think strategically about the conversation as a whole. Thinking more about a conversation overall provides confidence and smoothness that will help us act intuitively in the moment. The macro question is: What are our goals? While the micro question is: What topics are going to help us get there?

In practice, this means we need to rely a little more on our deliberative system 2 brain. For my students, thinking of a topic ahead of time, such as Dr. Manhattan in *Watchmen,* made their conversations better. While smoothness and excitement can happen spontaneously, too, they are actually more likely to happen with deliberation and forethought. This is true because of a phenomenon psychologists call "cognitive offloading," reducing the information-processing requirements of a task through physical actions (such as writing down ideas or taking notes) to reduce cognitive demand. It's extremely difficult to manage topics in mid-conversation, while you're trying to listen and respond and are feeling panicked about "oh my gosh, what are we going to talk about next?" It's much easier to brainstorm purpose-driven topics before a conversation begins, when we have time to engage our system 2 brain. Topic prep, in turn, frees up our mental space during the conversation, so we can be more attentive, feel less panicked, and speak more fluently.

When people come up with topics ahead of time, the quality of their conversation usually improves immensely. In studies I've worked on, telling participants to think ahead about possible topics *even for just thirty seconds* increased the fluidity of their conversations, especially at the often-awkward transition points between topics. Those who had prepped topics took shorter pauses and used fewer disfluencies like *um, uh,* and stutters. Privately, they *felt* smoother too—less anxious, more confident. And being a little smoother—just a little!—makes a big difference. Research shows that greater conversational fluency (or flow), especially at topic boundaries, improves networking success and feelings of chemistry and connection. We're able to build better relationships, in which everyone feels more comfortable and less awkward.

Topic forethought even improves the accuracy and relevance of the information we exchange. If there's something we need to remember to talk to someone about, we're more likely to remember if we think about it ahead of time. In addition to remembering what we need to discuss, prepping topics also makes us cover *more* topics, which can bring its own benefits, especially if things are moving too slowly and the participants are losing interest. Prepping topics makes switching topics easier and smoother (less stilted, gap-ridden, and redundant), so people end up switching topics more frequently, and they are more likely to alight on mutually enjoyable ones.

Always Thinking Ahead

Regardless of whether on-the-spot topic brainstorming comes to you naturally, forethought is your friend. Thinking ahead about topics can help us all.

I don't mean sporadic topic forethought once in a while, before especially important or high-stakes conversations. I recommend *habitual* topic forethought before *every* foreseeable interaction. I know—that's a lot to think about. But it works. Simply knowing that topic forethought is possible—and permissible and helpful—makes people more likely to think about their upcoming conversations and their partners, and to reflect on which steps will stave off unpleasant scenarios and spark pleasant ones. Topic forethought makes you think about your conversation partner, and as we'll see throughout this book, almost all the secrets to becoming a better conversationalist have to do with learning more about your partner's perspective.

In both my research and my teaching, I've found that despite initial skepticism, once people experience the benefits of topic forethought, they *really* get into it. They make topic notes to themselves in their Google calendars. They budget extra time to reflect before meetings. They develop spreadsheets filled with promising topics to raise with strangers. They know they don't *have* to raise their prepped topics during a conversation, but the topics are ready when they need them. If you get addicted to topic forethought, you won't be alone.

The Big Power of Small Talk

Among the many people who are missing out on the benefits of topic forethought are Hugo and Josie.[*] Hugo recently arrived in California from France—his English is quite good. Josie's in grad school at Stanford. They're on a four-minute speed date:

JOSIE: So, how do you like California?
HUGO: Wonderful.

---

[*] All the conversations in this book are real! But the names and some details about participants in the research studies and private conversations have been changed to protect their privacy.

JOSIE: Yeah?
HUGO: Beautiful.

Oof. One-word responses are tough. Josie asked an open-ended question, and Hugo basically gave her nothing to work with. Already running on fumes, Josie offers a few notes on places Hugo could visit—Palo Alto, San Francisco. Then Hugo switches to that most classic of all conversation standbys:

HUGO: The weather is incredible.
JOSIE: And this is even bad weather! It was much nicer last year.

Oh, no, you might think, *not the weather.* Yes, my friends, the weather. It goes on:

JOSIE: What's the weather where you're from in France right now?
HUGO: It's warmer than here.
JOSIE: Oh, really?
HUGO: But we have more rains. It's like Great Britain and—
JOSIE: Rainy all the time?
HUGO: In the south it's really hot. South, middle south, it's quite good weather.

You might be thinking: *This conversation is painful.* Even today, in a world of devastating tornadoes and hurricanes, snowstorms where and when they shouldn't be, and dry seasons so long and arid that whole regions seem to catch on fire for months at a time, few conversational topics have a worse reputation than the weather. It's the ultimate cliché, the epitome of dullness—an unacknowledged but glaring conversational failure on the part of everyone involved and any random stranger within earshot. Conversationally, the weather seems like the stuff of nightmares—a topic that condemns you to a seemingly endless loop of empty chitchat and filler, yields little, and seems to lead nowhere.

This, indeed, was the gloomy fate of our two speed daters. When the researcher asked Josie and Hugo to rate their conversation afterward, both gave it abysmal marks. There would be no second date.

Their exchange captures the particular torture many of us feel during small talk. Traffic patterns, your weekend, the baby's sleep schedule—these are among the topics that often leave us gasping for oxygen or desperately looking for an exit, even when both parties truly want to connect.

Most of us dread so-called small talk. Some people even tell you directly as soon as they meet you that they hate it. (And, by the way, bonding over a shared loathing of small talk can be one way to escape it, though not always.) There are lots of good reasons to feel this way, as our doomed speed daters can attest.

But hang on a minute. As much as Hugo and Josie's failed attempt at conversation seems to prove the point, I want to offer a counterpoint: small talk wasn't the problem here.

That conversation didn't go well for many reasons—among them the fact Hugo gave mostly one-word answers and asked few questions, problems that are hard to overcome. But the kiss of death was not the small talk itself—after all, they'd just met and knew nothing about each other. It was the duo's clumsy topic management. They got stuck in a small talk doom spiral, somehow unable to move from small talk to a higher level of conversation, even though everything they needed to do so was right there in front of them. Small talk is a doorway to something better, but neither Hugo nor Josie grabbed the knob to open the door and walk through it.

Contrary to popular belief, small talk is *crucial*. It is the initial proving ground, well trodden and generic, that often enables us to move on to bigger, deeper talk. It allows us to converse with strangers and new acquaintances whom we know little about—and to reconnect with people we know well. It's critical in mixed company, too, where people's preferences and relationships are all over the map. One person in the group may be dying to inquire about her friend's mother's recent stroke, while another would be more comfortable talking about bird-watching. Relatable patter can keep the conversation alive without making assumptions or favoring the interests of one person or another. The key is to think of small talk as a starting point, a means to an end rather than an end itself.

You can also think of small talk as part of a broader category: portable topics that you can talk about with *anyone*. Its universality is what makes it powerful. You can talk with anyone about the weather, or major current events, or food, or sleep, or whatever's happening in your immediate surroundings. Anyone will be able to engage with you on these common topics. But when you stop to think about it, you realize that these topics are common because they're right in front of us, the easiest to grab on to in the moment. Small talk's usual suspects are what our system 1 brain rounds up from the current environment. And yet any of them, managed well, can move a conversation somewhere better.

Here's where it gets more fun. These familiar examples are not the only universally relatable subjects—they're just a small subset. *Topics we can talk about with anyone* also include big, abstract topics that tap into the human experience—like "What have you been excited about lately?" or "What can we celebrate about you?" These topics, by virtue of their ability to grab a partner's attention and ego involvement, tend to catapult us away from small talk. If the weather and the meal you just ate are the bottom rungs of the conversation ladder, these topics may help you climb up out of the doldrums more quickly.

Shifting the way we think about small talk from system 1 to system 2 unlocks a world of possibilities. Suddenly, it's another area where topic forethought can pay dividends—by helping us keep more interesting, go-to topics in the wings and giving us the confidence that comes from knowing they're there. When you come armed with more compelling topics, you can skip—or skip past—traffic and the weather quickly.

Once you start brainstorming ideas to add to your stable of universal topics, there is no end of possibility. Some of my students have even chosen to build spreadsheets of what they deem to be good one-size-fits-all topics. I have habitual ones too, like "Who made you laugh recently?" and "Have you been listening to any good music?"*

We can draw inspiration from *everything*. Mike, the director of strategy and standards for a tech company based in Austin, Texas, and the husband of a colleague, told me he prepares topics by listening to *This American Life* during his daily run. Every episode of this public radio program turned podcast has a theme ("I Work Better on Deadline," "Day at the Beach," and "Bloody Feelings."). The host, Ira Glass, and a team of journalists craft stories from firsthand interviews centered on that week's topic.

But Mike—unlike (I assume) most of the other 3.5 million people who listen to the show every week—doesn't listen to *This American Life* just for his own edification. The show populates Mike's mind with an array of vivid stories, timely factoids, and incisive ideas that he can *talk about with others*.

The "House on Loon Lake" episode, for example, told about a group of pre-teen boys who found a mysterious, overgrown house in the forest that had been abandoned with salt and pepper shakers on the table, dishes in the sink, an invitation for a town hall dance, and a dress with ruffles hanging in the closet. Mike converted it into a brief, portable story. "Can I tell you about a story I heard recently on

---

* Of course, just because you can raise these topics with anyone doesn't mean you should raise them with everyone. Even relatable topics are sensitive to context. Take, for example, news headlines, which might seem like obvious fodder. Certainly, it is a good idea to pay attention to the big things happening in the world, so you can be ready to discuss them. But these days the news comes with a warning label: everyone receives different news from different sources, curated differently, with differing perspectives. These differences can be a rousing source of conversational fodder, or they can create a rift in our shared reality from the get-go.

NPR?" he'll ask a colleague (usually during a lull in technical shoptalk). He'll provide some quick details, then ask something like, "Why do you think they *left*?"*

The effectiveness of his *This American Life* habit has led to other types of topic prep. Mike went on to steal the questions from the end of the series *Inside the Actors Studio,* like "What is your favorite curse word?" and "What sound do you hate?" He doesn't hide the fact that he grabs these stories and openers from podcasts and TV shows. "Once you get into it," he says, "it doesn't matter where they came from." He's right. It matters where they *go.* And the world of topics, including small talk (yes, even the weather), is filled with possibility.

Hunting for Treasure

Mike's *This American Life* gambit shows the hidden power of small talk. But it doesn't totally solve the critical small talk problem—how to stave off a conversational doom loop. Why not? Because even the more interesting, broadly relatable topics can also trap people in the kind of soulless interpersonal death spirals that our Californian speed daters endured. No one really wants to talk forever about how they're sleeping or what happened at the house on Loon Lake. There's a limit—even when it comes to the sparkling topic choices concocted by the expert talkers on *This American Life.*

The trick is timing: Don't stay too long. Even with good small talk, you want to use it only as long as necessary. The key is to steer away before it starts to feel dutiful. To do that, as always, you must be a treasure hunter.

---

* Other episodes serve as unconventional conversation openers, too. The episode "Mapping" followed people who make maps by tracking untraditional landmarks, like cracks in city sidewalks or the homes of Hollywood stars. Mike will ask a conversation partner, "What would you want to map?" or "What if we mapped smells instead of streets?" That opener has led to conversations about how his colleagues see colors and how his wife's friends perceive reality.

Think of conversation as a vast ocean dotted with infinite topic islands. Great conversationalists are hungry to discover new terrain—the uncharted territory that poses risk but also offers the possibility of new rewards. Sailing peacefully through the calm, well-charted seas of small talk feels safe, and we can use that sense of safety to scope out promising islands. But it's easy to stay on the boat too long, admiring the islands from a distance. You'll never get any treasure that way. Eventually you have to take a risk—you need to go ashore and explore new ground.

Here's the secret upside of small talk: we can use it as a stepping-stone toward bigger, better, more interesting conversations. It's an exploratory space that helps us look for opportunities to climb toward more meaningful or interesting topics, especially with strangers or people with whom we share little history. Talk of the weather could lead to the discovery that a conversation partner was out of town, which could quickly lead to a lovely discussion of Tahiti. For Josie and Hugo, it wouldn't have required a dramatic leap to move from the weather in France to his childhood growing up in France. Or from the weather in California to that favorite shady bench Josie's discovered on campus, to childhood secrets, cultural differences, hopes and fears, music, or where and when they have their best ideas. Small talk, managed properly, is a well-worn ritual that gives our brains the time and opportunity to formulate better, deeper questions and ideas. "The merest everyday speech-morsel," writes *Atlantic* columnist James Parker, might all of a sudden "tip you headfirst into the blazing void" of another person's "soul."

Climbing, Up and Down

Rising from the mundane world of small talk to the blazing void of the soul takes some skill. Rarely does someone's comment about the light rain outside lead instantly to a deeper discussion of how a person's relationship with their mother is the key to understanding their life choices. There are lots of conversational steps in between, in which we figure out which topics can take us where we want to go.

It helps to organize our infinite topic options into a hierarchy—let's think of it as a *topic pyramid*—with the most broadly relatable topics at the bottom, and the most uniquely personalized at the top.

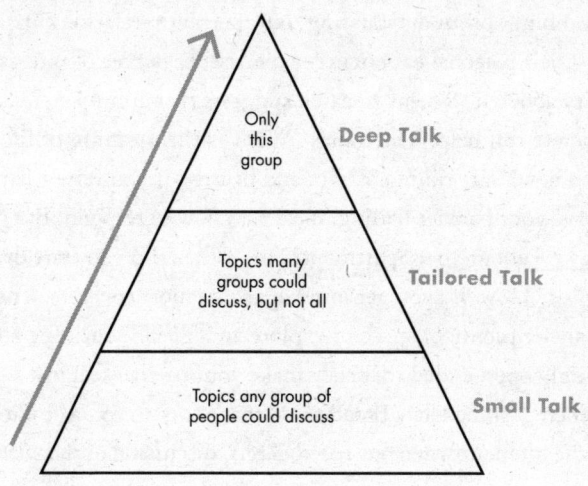

The topic pyramid: moving from small talk to deep talk, and back again

At the base of the topic pyramid are things everyone can talk about, labeled small talk. The base is often a useful place to start, and it can even be the right place to stay—in brief exchanges with baristas or neighbors, for example. But because these topics are depersonalized, lingering at the base of the pyramid can also feel impersonal, dissatisfying, shallow, or even dehumanizing.

One goal is to move up the topic pyramid, out of the base level, while realizing that the conversation may go through cycles from bottom to top (and back again). Some conversations may be better suited to tailored talk than to deep talk. We're not always headed for the peak. But we should have some idea of how high we and our partners might want to climb, and we should make sure that, no matter what, we don't get stuck for too long at the base.*

Making the leap from small talk to more tailored topics takes skill. But it's easier than we think to jump from the weather in France to growing up in France or the allure of French rap music, or from the weather in Italy to opera or motorcycles. Making these leaps has to do with probing or understanding our partner's relationship to the topic—their personal experience, expertise, or degree of interest in or curiosity about it. Recent research suggests that aiming to make topics concrete can help. The topics "What is the meaning of life?" and "What's new?" are quite abstract and broad—if you raise them, they may leave your partner flailing, uncertain how to respond. But "What do you love about your apartment?" or "When did you start drinking coffee?" or "How did you get into yoga?" are more concrete. They give your partner plenty of space to explore and ascend, but they aren't so dauntingly open-ended that they make your partner feel lost—you've given them a sufficiently broad topic that's easy to connect with.

At the moment when our speed daters' discussion of the weather in California and France was sputtering, Josie could have asked a more concrete, personal question, such as "Do you like California because it reminds you of where you're from?"

Follow-up questions are great for moving conversations in a more concrete direction (much more on them in the next chapter). But disclosures can work just as well. When Josie said "rainy all the time" with a hint of disappointment, Hugo could have disclosed something like "I actually sort of love it when it rains—I think it can be really cozy." Revealing this kind of personal information nudges the conversation gently up the topic pyramid, because personal disclosure tends

---

* As we'll see in Chapter 4, L Is for Levity, we can make the base of the pyramid feel more or less rewarding. The success of any topic is shaped by the emotion underlying it. The meaning of life can feel like small talk if discussed flatly, while talk of cold pancakes can be quite special or meaningful if discussed with fizz.

to trigger reciprocal disclosure from a partner, and because we move up the topic pyramid on the motor of vulnerability and trust.

A good conversationalist is always on the lookout for personal details, which are both concrete and ego-involving. If your talk partner has kids, you might talk about their kids or about children in general ("What are nine-year-olds into these days?"). If your talk partner works, you might ask about that ("What's been interesting to you at work recently?" or "How did you end up in this job?"). Commonalities are especially rich topics. In the Chat Circle exercise, the topics my students prepare—"Has Harvard Business School lived up to your expectations?" "What did you do over spring break?"—tend to take advantage of their scant knowledge about their partners: the fact that they too are a student at HBS embeds fertile common ground into that shared reality.

As I've noted, the top of the topic pyramid isn't the destination for every conversation. The middle of the pyramid has much to recommend it—it's a great place to learn more about someone, to get a sense of their tastes and preferences, and to let their personality shine through when talking about subjects that engage them. We don't need to get to the top with everyone, but some of the deepest human rewards lie there.

If we can navigate our way through small talk and tailored talk, we may reach the top of the pyramid, which includes things that you and your partner want to talk about—and can talk about effectively—with each other. You land on topics on which you and your partner can achieve your most important motives to connect, savor, protect, and advance. These are usually not topics that anyone could talk about; they're uniquely suited to a particular combination of people, to their particular common ground, at a particular moment, in pursuit of a particular set of goals. Maybe it's something you've discussed before to reify your relationship, or a secret your partner wants to share with you, or a specific problem or triumph that they're facing that you're positioned to receive and support. And for first-time talkers (strangers), whose goals often focus on getting to know each other and on not feeling awkward, getting to the peak can simply mean finding a topic on which either partner has something exciting, funny, or insightful to share.

For our speed daters, if Hugo admits how homesick he is, then Josie can share an insight that has helped her cope with being away from her family that might also help him. If Josie talks about a restaurant in town that she loves, Hugo can express his excitement to go there, too. If Hugo mentions a very specific indie band he's curious about, and she's curious about them too, maybe they can plan to see them together on their second date. Reaching the top tier doesn't mean your partner won't or can't talk about those topics with others. It just means they feel interested, happy, and safe enough to talk about them with you. Being able to find and discuss peak-of-the-pyramid topics speaks to the strength of your relationship, and discussing them strengthens it more.

Personalized Prep

Forethought—thinking ahead about the world of your partner—is key for reaching the top tier of the topic pyramid (deep talk). Once my students have bought into the idea of topic forethought, they start to ask themselves questions about the people they're likely to run into:

- What happened during our last conversation that I can call back to?
- What's happened in their life since we last spoke that I can or should remember to ask about? Nearly anything will do: "You were just about to start guitar lessons. How's that going?"
- What might be on their mind now that would be fun, productive, or important to discuss?

> - What can *I* share about my perspective or my life that might be interesting, fun, or helpful to them?
> - How can I keep track of things in the world that remind me of this person—like news articles, memes, YouTube videos, TV shows, headlines from *The Onion,* whatever—so I will remember to raise them as topics when I see them?

Let's Move On

After Hugo and Josie reached their eighth conversational turn in a row, still idling over the weather, it's not hard for us to imagine how they were feeling. *Shudder.* And then there's a conversation I overheard at the office recently:

"How about M&M's?"
"Yeah, I like those."
"Reese's Pieces?"
"I'm allergic to peanut butter."
"Almond Joy?"
"Is that the coconut one with or without nuts?"
"Mounds don't have nuts."
"Oh yeah, I don't think I've had either one."

A passing mid-October comment about loving Halloween—a topic brimming with promise and potential pathways up the pyramid—got stuck in peanut butter and chocolate. Maybe the people in this conversation missed the chance to veer into talking about the costumes they preferred as kids, or whether they really believed in ghosts, or the harrowing history of the Salem witch trials, which would have been more fruitful. But here they are, trapped in the candy aisle of small talk:

"Twix?"

"Only if they're fresh. Otherwise the caramel gets stuck in my teeth."

"Yeah."

You probably know that familiar feeling when a conversation is on the brink—you're confident that all the good juice has been squeezed from that orange, and everyone's barely subsisting on pulp and seeds. You may suspect this fragile state of affairs is common knowledge, that your conversation partners are not only aware that the conversation is stagnating but also aware that *you* are aware that *they* are aware, and so on. Or perhaps you suspect that *no one* else realizes how malnourished or misguided the conversation has become, and it's clear that you need to deliver a juicy new orange *pronto*.

So far we've covered the *what* of topics, but the *when* is just as important. A topic's sweet spot is a little like a fruit's ripeness cycle. It's not there . . . it's not there . . . it's not there . . . then for a hot second it's perfect . . . and then it's gone on too long. The timing and the duration of conversation topics have just as much impact on our feelings about the conversation as the topics themselves. Conversations don't last forever, and we have to make choices about which motives to pursue, via which topics, and for how long. When a topic stagnates, a conversation becomes malnourished, and our interest and engagement flag. And when our engagement flags, we can't pursue our goals effectively.

But when should we switch topics, and how? In the best-case scenarios, switching topics is seamless—they're easy off-ramps rather than abrupt hairpin turns. Sometimes both people realize a topic has run out of juice, but neither knows how to get to the next one (the birth of the uncomfortable silence). It's worse when one person feels a subject has run out but their partner doesn't. And while we may get an intuitive feeling that we need to change topics, we're not always great at picking up that feeling in our partners.

When a topic runs out of juice, one solution is simply to switch topics more often. After all, a subject can't run out of juice if you switch from it when it still has some juice left in it. My fellow researchers and I ran a study using pairs of strangers: we told some of them to switch topics very frequently—to try to get through twelve assigned topics in ten minutes—and we told other pairs to move through the same list of topics naturally, at whatever pace they wanted. The question we were trying to answer was: What leads to better conversations—switching more frequently to allow for the discussion of more topics (even if it means leaving some juice behind), or just letting conversations take their natural flow?

Lo and behold, our studies found that people who switched topics more frequently enjoyed their conversations significantly more. On average, frequent switchers rated their conversations a 6 out of 7 in terms of enjoyment, while natural switchers rated theirs a 5 out of 7. Compounded over many conversations, a 1-point difference out of 7 is a really big deal. Do you want to live a 5-out-of-7 life or a 6-out-of-7 life?

The association between topic-switching and conversation enjoyment applies in real life, too. In a dataset of more than 1,600 thirty-minute conversations online, most people felt they covered "the right amount" of topics during their conversations, but more people felt they had discussed too few topics (20 percent) rather than too many (11 percent), and the negative impact of discussing too few topics on enjoyment was significantly higher than that of discussing too many. The good news is that the "too few topics" worst-case scenario is easily avoidable. In our switch-more experiments, we found that talkers could easily switch topics more frequently when they tried to: pairs who were instructed to switch topics more frequently covered an average of nine topics in ten minutes, while natural-switching pairs covered only five topics. When people are encouraged to switch topics more frequently than they normally do, they are able to do so easily, and it improves their conversational experience.

But here we have a problem. If we are trying to ascend the topic pyramid and get deeper into the topics that only a particular group could discuss, how can we make it there if we keep switching topics? Does a broad range come at the expense of the depth we're seeking?

Reader, it doesn't. My fellow researchers and I were surprised to learn that the folks who switched topics frequently—and achieved broader subject matter as a result—didn't do so at the expense of depth, a trade-off some of us worry about. Our research shows that those who manage topics deftly dig deep, find treasure quickly, *and* switch to new topics seamlessly, without painfully awkward gaps or delays.

So if switching topics leads to better conversations, why don't we all switch more often? You might say, "Because we didn't know about it and you just told us." Okay, fair point. But there are also more intrinsic reasons. Our default system 1 thinking is at least partially to blame. Our intuitive minds lead us to choose topics that are easily accessible, to omit less obvious topics that would be great to discuss, to switch away too early from good ones, and to linger too long on others. But it's also a matter of faulty mind-reading. How good are we at figuring out when our partner thinks a topic has run out of juice? Are we missing valuable clues that our partners are giving us about their waxing and waning interest in the topics we're discussing?

Noticing Others' (Dis)Interest

My research team wanted to understand how well people read their partners' cues when it comes to their level of interest in a topic under discussion—to see how well people weigh various signals of their partner's engagement. We found that people tend to weigh some cues, like follow-up questions, correctly. However, people tend to overfocus on some of their partner's behaviors and underfocus on others. Suppose you ask your partner how their parents are doing, and they respond, then come back with the mirror question—"How are *your* parents?" You might weigh that question too heavily, taking it as a sign that your partner is interested in that topic when actually they're not. Often, mirror questions are just the result of a reciprocity reflex. On the other hand, when a partner calls back to a topic you discussed earlier, it's a reliable sign that they like the topic and want to stay with it.

At the bare minimum, live conversation shouldn't feel like a monologue delivered to a politely muted audience. That doesn't mean there's no room for storytelling, or that parts of a conversation won't involve one speaker taking more airtime. But when a chat is going well, topics and stories tend to be "co-narrated"—listeners interject, ask questions, laugh, and give back-channel feedback like *yeah, uh-huh,* and *no way!*

We need to pay more attention to these signals of co-narration, because they turn out to be important indicators of a partner's interest. In their absence, you're likely to just be hogging airtime, without mutual interest and involvement. If either party's engagement is flagging, it's time to switch topics.

There's a final reason for our imperfect topic-switching: politeness. Sometimes we recognize the cues telling us to switch, but we ignore them because we don't want to be rude, abrasive, or otherwise jarring. I suspect this played a role during our speed daters' stilted conversation. Josie may have been dying to hear about Hugo's childhood, but she was also afraid to veer too sharply away from the weather for fear of seeming abrupt or intrusive. *Très gauche, non?* No! Most people are generally too cautious about topic-switching, for lovely but misguided reasons. It feels safer and more respectful to keep marching straight ahead.

The good news is that you can get better at figuring your partners out if you know what to look for. If you're not sure when to switch to a new topic, then you can look for three reliable, practical signs that it's time to switch—that the topic at hand is stagnating, the fruit is running out of juice:

- Longer mutual silences (pauses)
- More frequent polite laughter (a gut reaction to fill the longer pauses)
- Redundancies (people repeat things they've already said)

I imagine that Virginia Woolf and Stephen Spender's notably lackluster lunch was chock-full of all three of these telltale signs. But the presence of even one alone indicates it's time to switch. Other signs reliably indicate that a topic might be going better than you think: a partner laughing genuinely at his own words means he's amused, and the fact that he raised or called back to the topic to begin with means he wants to discuss it.

To be fair, keeping an eye out for cues that it's time to switch topics seems like a lot of work. You might be wondering, *How do I do it? How do I think about what to say, then say it, then listen to the response, process the words, and also pay attention to subtle cues about how my partner is responding?*

By this point, I hope the answer no longer surprises you (or my skeptical student Josh): forethought! When it comes to reading a partner's nuanced behavioral cues (mind-reading), the computational limits of the human mind put us at a major disadvantage: our brains can only handle so much. That's yet another reason it's so powerful to think ahead—doing so frees up our mental space, allowing us to read our partner's unspoken cues. If your partner raised the topic and is still calling back to it, that's a signal for "don't switch." But if either of you is starting to repeat yourselves, that's a signal for "time to switch." You can also do your partner a great favor—sparing them some mind-reading work—by telling them when you are or are not enjoying a topic. A little feedback can help a lot.

Master conversationalists—people who are especially successful in my studies of topic management—think ahead about what they and their partner want to get out of a conversation, choose topics to pursue those motives, and don't shy away from changing the plan when topics lose their juice. In doing so, they achieve both breadth *and* depth. Most of us, if we can think ahead a little more and switch topics a little more confidently, will find that we can also "create an atmosphere" like Primo Levi's Super Greek.

Making Topics Good

One day a few years ago, my college friend Carolina called me. Carolina is fearless. She doesn't hesitate to raise topics that would make many people squirm. In just one month, she initiated group text chains with the following:

> "Is it irresponsible that Elizabeth Holmes had a baby during her trial? It seems like a ploy for sympathy."
> "I'm bothered by people who care about the safety of nonstick pans."
> "Has feminism, overall, been net positive or negative?"

Carolina called me during the lockdown period of the pandemic. She was at home in Oregon, while I was outside for an escapist walk around my block. I answered before the second ring.

"Hey!"
"What do you know about vaginal mesh?"

I laughed out loud. As a mother of three, I knew a few women who had needed mesh to repair their pelvic organs after giving birth, and I'd heard something about a class action lawsuit related to vaginal mesh, in . . . Australia, maybe?

No question that for many, this opener would be a turnoff—brazen, sort of gross, and possibly a gateway to a world of stress, complaints, and intrusive (or just plain boring) chitchat about childbearing. But I wasn't turned off or grossed out—far from it. I understand the physical toll of pregnancy and childbirth, and I love Carolina's racy topic-steering ways. It was so *her* to start a conversation that way.

"Ha! I'm so happy to hear your voice," I said. It was true. Her presence, even over the phone, was a balm.

"You too, Al! I need your help. I have to decide if I'm going to repair my lady parts now or have another baby first."

I asked Carolina how she was feeling about going for child number three. She wasn't sure. She'd always wanted a big family, and she said she couldn't imagine not meeting her next child, but she'd also just switched careers. She'd had her second child while she was in business school. Having a third one, during her new job, would be tough.

She made a very quick topic shift from vaginal mesh, with its concretely visceral shock value, to a sincere plea for family planning advice—a shift that led us into an honest discussion about the challenges and rewards of working parenthood. These were topics we both needed to discuss, for comfort, for support, for escape.

We talked about the relentless, maddening effort to convince two small humans to put their shoes on while you're struggling to slip on an even smaller human's sneakers. Then our conversation wound its way to dodging fingerpaint handprints on a suit, to the intoxicating joy of lounging around with a sleeping baby in your arms, to the profound elation when your oldest kid makes a really good joke for the first time in their life.

We moved on to our marriages and vacations and pandemic struggles. Then, as I rounded the last corner of my block, she spilled, "I'm having some troubles with my mom and my brother."

Of all the things Carolina had surprised me with over the years, this was probably the most unexpected. Her father had died when she was eighteen, just before we met at college, and she'd always been very close with her mother and brother. She started to cry as she explained that they felt very far away, and they were having a hard time forgiving her for moving to Oregon, and the pandemic had made it all so much worse. It was heart-wrenching to hear, but I felt deeply grateful that she shared it with me. I told her it was okay if her family strife was muddling her decision about whether to have another child. She said, "Thanks, Al. Love you."

And then, back at my front door, with my own kids tumbling around inside, I had to go. Separated by decades and thousands of miles, I still felt so close to Carolina. Good conversation doesn't always require that we choose good topics. Even the most ridiculous topics can become deeply meaningful. If you and your partner pursue goals that really matter to each other, every once in a while you'll land on something that's truly revealing, real, and rewarding (especially if you think ahead, as I suspect Carolina did before picking up the phone to seek my advice).

A little over a year after our call, Carolina welcomed her third child into the world. She posted a few photos of the baby online. There was no mention of vaginal mesh, but her mother and brother were beaming by her side.

THREE KEY TAKEAWAYS FROM CHAPTER 2

T Is for Topics

- **Topics are the building blocks** of conversation.
- **Small talk isn't the enemy.** Getting stuck on any one topic for too long, especially topics at the base of the topic pyramid, is the enemy.
- **Topic prep** is your best friend.

CHAPTER 3

A Is for Asking

CARRIE FISHER WAS TWO YEARS old when her parents separated—and it became the biggest tabloid scandal of the day. Her mother was Debbie Reynolds, who starred in *Singin' in the Rain;* her father was the pop singer Eddie Fisher. For years, her parents had been close with the Hollywood icon Elizabeth Taylor. Reynolds and Taylor had even been childhood friends. When Taylor's husband died in a plane crash in 1958, Fisher rushed to her side to comfort her. Then he decided to stay there, leaving Reynolds to raise their two young kids, Todd and Carrie, by herself.

Carrie entered the family business as a teenager, playing a supporting role to her mother in the Broadway musical *Irene.* When she was in her twenties, her portrayal of Princess Leia in the *Star Wars* movies launched her to superstardom. Alongside her fame, Fisher struggled with bipolar disorder and drug abuse. In the mid-1980s, she nearly overdosed on tranquilizers and had to check into rehab. Later, she admitted that she had taken cocaine on the set of *Star Wars: The Empire Strikes Back.* Her reputation tanked, and her acting career floundered. But by the end of the decade, she was getting better roles and was turning to her own tumultuous life for artistic inspiration. She wrote a novel, *Postcards from the Edge,* about a mother and daughter in show business—it was based heavily on her own life. In 1990, she turned the book into a successful screenplay and wrote another semi-autobiographical novel, *Surrender the Pink*.

That year, Fisher appeared as a guest on the program *Fresh Air* on National Public Radio. The executive producer, Terry Gross, had been hosting the daily program for fifteen years, starting two years before Fisher became Princess Leia. Gross was now thirty-nine, almost six years older than Fisher. In the interview, Gross and Fisher discussed Fisher's new movie, her new book, her relationship with her mother, and her struggles with mental health and drugs. Listening to the recording today, it's clear that the two are skilled conversationalists, not least in topic management. In their deft hands, even topics that some might consider sensitive or offensive aren't off-limits.

Once, after a relatively innocuous question about the types of roles Fisher gets, Gross seems to sense that she has permission to take Fisher to the top of the topic pyramid.

"Did nearly OD-ing and then recovering from your drug habit change the kind of acting you wanted to do, or the way people perceived you who were casting you?" Gross asks.

It's a jarringly intimate question, but Gross delivers it with warmth and a disarming directness. Fisher responds in kind, describing how her bad reputation lingers in the movie business. Gross acknowledges Fisher's disclosure with an understated "Mhm," then, to avoid lingering too long on a heavy topic, steps gracefully down the topic pyramid by asking about Fisher's early career. Fisher makes wry jokes, and Gross laughs easily and generously. To listeners, such a conversation between two pros makes for great entertainment. And that's why they're there—the more interesting the conversation is for listeners, the more Gross can burnish her reputation as an interviewer and grow her audience, and the more Fisher can entice people to buy her new book.

But even in this public context, something more elevated and perhaps intimate seemed to transpire between them. The interview didn't have to go as well as it did, and one reason it did was Gross's skillful questioning. Her questions guided the topics, unearthed new revelations, and seemingly brought Fisher and herself closer together. Each one was an invitation for Fisher to make herself known. Why didn't you star in your own film? Could you elaborate on that? What kinds of things did you and your mother fight about? How did being photographed as a child affect how you feel about being on camera? Do you look back at those pictures now? No but really, do you have them anywhere? What was your audition like for *Star Wars*? Do you ever regret dropping out of school? Do you sing anymore? How has Hollywood changed since your mother's time? The questions provided a brilliant skeleton for the conversation, a skeleton fleshed out by Fisher's responses.

This was the first of three interviews, over nearly three decades, between Gross and Fisher. The next one came in 2004, the third in 2016. The three interviews trace the arc of Fisher's life from age thirty-three to age sixty, exploring with ever more detail the themes that had defined Fisher's life: Hollywood, her famous parents, *Star Wars,* feminism, marriage, divorce, substance abuse, movies, writing, and mental illness. At the same time, the interviews charted the growth of Fisher's relationship with Gross, which became closer and more comfortable with each new conversation.

In 2016, Fisher published her third memoir, *The Princess Diarist,* based on the diary she had kept during the filming of the first *Star Wars* movie. Fisher had forgotten about the diary until, decades later, she found it in a box beneath her floorboards. *The Princess Diarist* described the early days of *Star Wars* and recounted Fisher's experience with sudden stardom. But by far its biggest revelation, the revelation that made it into every headline about the book, was that Fisher had engaged in an affair on set with her co-star Harrison Ford. Fisher was nineteen years old at the time. Ford was thirty-three and married, with two kids.

In November, Fisher went on *Fresh Air* again to promote the book. It was one of her last public appearances. Almost exactly a month after the interview aired, she passed away from a heart attack.

Early in the interview, Gross addresses what must have been the elephant in the room. "So what really made news from your book is your affair with Harrison Ford when you were making *Star Wars*," Gross says. "Did you tell him you were going to write about it before you actually published the book?"

"Oh, yeah," Fisher says.

"I'm really relieved to hear that." Gross laughs. Fisher reassures her that she wouldn't have published the memoir without letting Ford know.

"But it's still, no matter if I told him or not, it probably feels like an ambush. It feels like an ambush to me, and I'm the one that wrote it," Fisher says.

Gross probes more deeply: "Did you tell him or did you ask him for permission?"

"No. I said, 'I found the journals that I kept during the first movie, and I'm probably going to publish them.' And he just sort of raised his finger and said, 'Lawyer!'" Fisher says. Gross laughs. Anyone who knows Ford's gruff sense of humor can imagine the exchange. Fisher explains that she sent the book to him before publishing it and gave him the chance to take anything out that he wanted to—but she never heard back.

Fisher reads a passage from the original diary, then Gross follows up on the question of her affair, pressing for richer insights. "Your relationship with Han Solo, Harrison Ford in the movie, is such the kind of, you know, typical will they or won't they," she says. "They hate each other, but that's because they really like each other, you know? And so you're having that kind of onscreen relationship, and in real life you're having an affair. So how did the affair affect the chemistry on screen?"

"I think it made us more comfortable with one another," Fisher replies. "I think it made me more able to wisecrack to him. Even if I was insecure, we were having an affair, so there was something to base some security on. I don't know." On the recording, she sounds as if she's working through her thoughts on the spot, as if Gross's question has sparked some new idea. "Well, there was chemistry there, and you can see it," she says. "So I don't know which came first, the chemistry in the film or the chemistry in the world."

Though Fisher had written about the affair with Ford in her memoir, Gross's questioning during the interview shed new light on one of the most epic onscreen love stories of all time. This question seemed to unburden the conversation slightly. Now that the pressing business is taken care of, they could just talk.

As we listen, a mood of nostalgia colors the rest of the conversation. They laugh about Fisher's overwrought teenage emotions, clumsy *Star Wars* dialogue, and Princess Leia's infamous hairstyle. In the three decades since their first interview, Fisher's voice has become raspier, and her thoughts seem to turn more to the past. Gross's voice, meanwhile, has become deeper and warmer. It's easy to imagine them as two old friends, reminiscing about ancient history.

But near the end of the conversation, Gross asks a question that takes them to a topic they haven't discussed in any of their previous interviews—Fisher's dog, Gary. "What kind of dog is he?" Gross asks.

"He is a French bulldog," Fisher replies. It turns out that Gary is in the studio with Fisher—she finds him soothing and takes him everywhere. "He's licking my hand right now. He's just very nice to have around."

"Oh my god, I hear him licking your hand," Gross says suddenly. She laughs. On the recording, Gary's rapid, wet licks are very audible.

"That is such a loud lick," Gross says.

"Well, he has a very big tongue," Fisher says. They both laugh. They chat for a while longer about Gary, guided by Gross's questions—he's a certified therapy dog, she found him at a "very tragic pet store" in New York, he's very well behaved.

"I can't believe I still hear him licking you," Gross says after a while. Soon, they wrap up the conversation. As they say goodbye, Gross gives her regards to Gary.

"I'll lick him for you," Fisher says. It's a fittingly joyful end to their thirty years of great conversation.

Moments like this are a testament to Gross's immense skill as an interviewer. Thanks to her meticulous preparation before each interview, her instinct for improvising discerning questions on the spot, and her decades of interviewing experience, she makes her guests feel safe expressing themselves honestly. But it's also a testament to the power of questions themselves to produce rich and marvelous conversations.

A Is for Asking

In the coordination game, if topics are the game pieces, then questions are the hands that lift those pieces, roll them between their fingers, and push them around the board. Questions are the most powerful tool we have to raise, switch, and stay on topics. Asking rises to the level of a maxim because without questions, conversations risk unfolding as parallel monologues between speakers. Questions, on the other hand, allow speakers to respond, interact, and build—they allow people to collaborate and co-create. Asking questions sets your partner up to disclose new information, which is the clearest path toward understanding their mind. Disclosure, in turn, can raise new questions so good conversation can flow back and forth, cascading between question askers and question answerers. The conversations between Gross and Fisher were successful because Gross's questions drew out Fisher's perspective. Questions are the best-known way to learn about others' perspectives—the necessary precursor to achieve our motives in every quadrant of the conversational compass.

Luckily, the ability to ask questions is a singular evolutionary gift. Captive primates, like bonobos, have learned to communicate with humans surprisingly well using symbols called lexigrams that represent human words. They can answer questions and identify objects. But even the most highly trained primates are unable to ask basic questions. Meanwhile, all across the world, human children ask their first questions in their babbling months, initiating requests for milk, food, and objects by pointing and making sounds with that recognizable upward-turning questioning intonation, long before they start using phrases or sentences. Even if only to ask for help or permission, question-asking is a fundamentally human approach to conversation. It reflects our interest in others' minds. It's what makes human conversation possible—and remarkable.

Ask More

People don't always think of the role questions will play in a conversation, perhaps because they sense that they need to know what to say—they think they need to be knowledgeable and interesting and assertive. But in my research, I've seen that asking more questions—during speed dates, sales calls, parole hearings, entrepreneurial pitches, and job interviews—correlates with all kinds of positive outcomes. The most obvious benefit is that question askers learn more information. When you ask, people answer, and you know something you didn't know before. Terry Gross, for instance, learned Harrison Ford's reaction to Fisher's planned memoir ("Lawyer!") by pressing Fisher with a question ("Did you tell him or did you ask him for permission?").

Asking more questions increases information exchange, but it also has a less obvious, and more important, benefit: it improves the relationship. People who ask more questions are *better liked*.

In one study, my fellow researchers and I brought together pairs of strangers and asked them to get to know each other in fifteen-minute conversations. We observed their natural question-asking behavior. Each talker tended to ask 6.5 questions on average. Then we ran an experiment with a separate group of strangers. Before they began their conversations, we gave them secret instructions. For half the pairs, we told one talker to ask a lot of questions (more than nine questions in fifteen minutes), and for the other half, we told one talker to ask very few questions (fewer than four questions in fifteen minutes). We didn't tell them what kinds of questions to ask or give them any other instructions—they could talk about whatever they wanted, however they wanted, to maximize enjoyment.

Those who asked a lot of questions, we found, were significantly better liked by their partner than were those who asked few. Asking more questions *caused* an increase in likability.

We've seen this effect in other places, too. On heterosexual speed dates, daters who asked just one more question on each of twenty dates were able to convert one more of those dates into a second date. One additional question per date meant a 5 percent greater likelihood of attracting a mate. And though men are much more likely to agree to second dates overall, asking more questions was equally effective for male and female daters. Can you imagine what success they would have if, in advance of the date, they spent as much time brainstorming questions as they did making reservations, choosing their outfits, or Googling each other?

The Dreaded ZQ

Even though question-asking increases learning, enjoyment, and likability, most people don't do nearly enough of it, research shows. Even in contexts that are theoretically designed for probing for information—meetings, date nights, job interviews, office hours—people often ask very few questions.

Professional matchmaker Rachel Greenwald calls the worst offenders "ZQs"—those who ask Zero Questions. Rachel's seen it all. While I sometimes study daters, and most people have been on dates, Rachel makes her livelihood from arranging, studying, and mentoring thousands of daters. But you don't need to be a professional to understand the problem with ZQs, she says. We've all encountered them:

the woman who talks endlessly about her kids; the old man who perpetually tells stories about his life; the boss who runs a meeting just to talk at his employees; the date who doesn't so much as ask, "How was your day?" At one point during these conversations, it may dawn on you that the person hasn't asked you a single question, and then the conversation becomes a potentially frustrating or distracting game to see if they will. In Rachel's words, "People say curiosity killed the cat. But when it comes to dating, curiosity is king: it's the ZQs that killed the date. Zero questions means zero second dates."

Thankfully, most of us aren't ZQs. But don't breathe easy just yet—we're probably closer than we realize. Whether chatting with friends, going on dates, or negotiating at work, people vastly *overestimate* how many questions they've asked during their conversations. Negotiators, for example, estimated that over half of their turns—more than 50 percent of the times they spoke—included a question. In reality, less than 10 percent of their turns included a question—less than one-fifth of the amount they thought they'd asked. We've found the same pattern in conversations between friends and on first dates, too.

Even when we do ask a lot of questions, we don't often notice their immense benefits. When my fellow researchers and I asked people to estimate how much their partner liked them in a follow-up survey, the people who had asked lots of questions didn't think they were liked more than those who had asked few. And it's not just because they were distracted by conversing. Outsiders who read the transcripts of other people's conversations didn't see the link between question-asking and its rewards, either. In the chaos of conversation, it's hard to see concrete links between micro-decisions and their outcomes—between asking questions and learning things or being liked. The invisibility of these links helps explain why people ask too few questions: they don't notice how much people like being asked abundant questions (and dislike being asked too few). They're not getting clear feedback.

But being asked questions feels good—really good. It makes you feel like the asker wants to know about you, to pull out the best ideas from your mind. You feel they care about you, and you want to see them again. It's no surprise that Carrie Fisher appeared on *Fresh Air* three times—Terry Gross is one of the best question askers out there, which encourages most of her guests to return.

But while Gross can serve as an inspiration, we don't need to be as phenomenal at asking questions as she is. To start improving our everyday conversations, we just need to *ask more* questions. They don't have to be dazzling, brilliant, or incisive. They just need to be present.

Sensitive Questions

Even in conversations where we want to ask questions, and we know it would be good to do so, we often feel afraid to venture one. We worry that posing certain questions will seem too sensitive or intrusive. But are people's fears of asking sensitive questions well founded in general? To explore this issue, a group of behavioral scientists compared people's predictions about how their questions would be received to reports of how they actually were received.

First, they crowdsourced a list of questions that people would like to know about a co-worker; then they ranked the questions from most to least sensitive. The least sensitive questions included "Are you a morning person?" "What season do you like best?" and "What are your views on pop music?" The most sensitive questions included "Have you ever had an affair?" "What are your views on abortion?" "How much is your salary?" and "Have you had sexual thoughts about someone of the same sex?"

Next, the researchers asked people if they were willing to ask their co-workers some of the most and least sensitive questions—yes or no—and how comfortable they thought their co-workers would feel answering them, on a scale from 1 to 7. The results confirmed that people tend to be unwilling to ask questions they think people will feel uncomfortable answering. The participants were twice as willing to ask the question anonymously (70 percent) as those who were not (35 percent), but even anonymously, they were significantly less willing to ask a question if they thought it would make their conversation partner feel uncomfortable.

But here's the million-dollar question: Would their partners *actually* have felt uncomfortable? Would those conversational moments be as cringe-worthy as they expected?

To find out, the researchers brought together pairs of people—some of them already friends, some of them strangers—to chat. In each pair, they randomly assigned one person to ask either sensitive questions or benign questions. After the conversations, they asked each partner to rate the conversation and the asker.

Between friends, people's expectations were dramatically miscalibrated. Asking sensitive questions didn't just have fewer negative consequences than expected—it didn't have *any* negative consequences. Recipients of sensitive questions like "How much is your salary?" or "Have you ever had an affair?" liked the conversation and their partners just as much as they did when they received benign questions like "How do you get to work?" Sensitive questions didn't cause any hitches in the conversation either—their recipients laughed, hesitated, changed the phrasing, responded, and asked follow-up questions just as much as the others.

So friends, it seems, can handle sensitive questions better than we think. But what about strangers? Again, they were wrong. Asking sensitive questions had no negative consequences among strangers either, whether these conversations happened in person or by text message. In five experiments with over fourteen hundred people, the researchers never found any evidence that asking sensitive questions is more dangerous than asking benign questions. Questions that seem sensitive aren't actually received that way—a fact we can't learn if we don't ask.

Of course, it's not that sensitive questions *never* have any negative impact. It's that they don't have a negative impact nearly as often as we think—and even when they do, the negative impact tends not to be as severe as we might imagine.

Reading the Room

But as always, success depends on context. Some circumstances make sensitive questions riskier in terms of damaging the relationship, so it makes sense to be more careful. For example, sensitive questions are more dangerous when the talkers are on a shaky footing—when they have broken each other's trust before; when they are in the heat of a serious disagreement or conflict; when they have power differentials that make the questions feel coercive; or when they are in large groups, where the desire for privacy and likelihood of shame are amplified.

Consider the famously nerve-racking question "Are you pregnant?" I consider myself lucky to have had a front-row seat at the "Are you pregnant?" question-asking show for almost three years over the course of three pregnancies. What a riot!

Women are pregnant for nine months on average—280 days, or forty weeks. But they don't *look* obviously pregnant for that whole time. During my pregnancies, people's fear of asking me "Are you pregnant?" shifted observably over time. Early in the pregnancy, a time when doctors describe the female body as "thickening," a question about whether you're pregnant can seem like an unflattering observation akin to "You look fat." Consequently, those who suspected but didn't know me well were afraid to ask, while those who knew me well did happy dances, gave hugs, and brought me bottled water behind closed doors. As my pregnancy became more obvious, I sometimes found it odd when people *didn't* ask about it, especially when we had nothing else to talk about—their question-asking reluctance became overt avoidance, so palpable that it was awkward. By month eight or nine, people seemed afraid to *not* ask. I was peppered constantly with questions like "How are you feeling?" and "When are you due?" It was fascinating to observe people's question-asking fears evolving over that nine-month period.

Most of us are afraid of asking questions that might appear intrusive. We worry that our questions will offend the recipient or reveal our own false assumptions or lack of knowledge—and so we opt to remain silent rather than risk causing offense. From my perspective, though, people felt more afraid of asking if I was pregnant than they needed to be. Boldly asking questions that feel risky can be a signal of intimacy—that a person feels comfortable enough to inquire.

Then again, on rare occasions I felt that the question "Are you pregnant?" was taboo for a particular reason. Early in my first pregnancy, I attended a barbecue with a number of work colleagues. A brash woman shouted across a room full of about thirty colleagues, "Why aren't you drinking, Alison? Are you *pregnant*?" I laughed and, a bit flustered, deflected with a joke ("Just trying to clean up my life!"). I was embarrassed to have my private business broadcast to a room full of co-workers, including some who would need to help me figure out the implications of my maternity leave—not to mention anyone who might have tried or might be trying to have children and couldn't.

I wouldn't have minded if she'd asked me whether I was pregnant privately—or texted a subtle "Are congratulations in order?" later. At the same party, a close friend noticed I wasn't drinking alcohol and asked quietly if I was pregnant. I was excited to tell her that I was. The question itself wasn't a problem, but the context in which the brash woman asked it was. Still, this experience didn't ruin my relationship with her, either. In fact, my memory of this story was harder for me to conjure than my memories of many quieter, joyful inquiries I was asked in the wings. Fleeting moments of awkwardness or tension in conversation tend to pass quicker than we fear and are forgotten more easily.

The benefits of asking sensitive questions often outweigh the risks because ascending to the peak of the topic pyramid—where we converse about things that reflect the unique reality shared between us and our partner—often requires that we ask sensitive questions. Our doomed speed daters Josie and Hugo would have been well served to jump away from small talk of the weather with any number of slightly racier questions.

We need to get deep and specific to reach scintillating topics. The very definition of *personalized* at the peak of the pyramid means we have to get *personal*—by sharing information about our lives and feelings; our unique knowledge (or lack of knowledge); and our preferences, work, opinions, and ideas. And getting personal makes us vulnerable. It opens us, potentially, to judgment and exploitation. But it also opens us to trust, enjoyment, and love—to feeling truly understood, respected, and valued. Though we may trip or stumble every once in a while, we shouldn't let our fears of making our partners uncomfortable be a roadblock that prevents us from truly connecting.

Asking Well

At the start of my Negotiation class (a course I taught before TALK), I used to hold up a hundred-dollar bill and ask my students to compete with each other for it. Who can convince me to give them the crisp bill? I'd been teaching them about social influence—how to use the powers of persuasion, while squarely avoiding threats, incentives, and deception. The lesson started with them trying to sell each other on the benefits of bacon-scented soap ("Perfect for vegetarians who miss the smell of meat," "Your dating life will never be better," "You'll be the star of the dog park"). We then went through several other exercises, exploring tactics like the power of extreme (but justifiable) opening offers, the power of loss aversion (people don't want to lose what they already own), the power of making concessions to trigger reciprocity, the power of aggregating losses and separating gains, and the power of social proof (when lots of people want something, everyone wants it more). But I'd saved my favorite demonstration for last. How will they convince me to hand over the hundred-dollar bill?

They wave their hands in the air shamelessly. They make stunning logical arguments; they promise me flowers, babysitting, and research assistance. They tell me they'll donate the money to someone who needs it more. They say they'll invest it on behalf of the whole class and pay dividends later. Like any good marketplace, the classroom is buzzing with noise and energy—with gasps and laughter every time a new suggestion is made. I stand there in silence, listening, still clutching the hundred-dollar bill.

Finally, a soft-spoken student raises her hand and says, "Professor Brooks, may I please have it?"

At that point, I smile and walk over to her, slowly. I kneel and bow my head as I hand the money to newly appointed question-asking royalty. While most of the students laugh and applaud, some yell *"What?!!"* or *"Come on!"* They're jealous—outraged by how simple it was all along.*

Some years the students don't solve the puzzle. Like the matchmaker's unattractive ZQ daters, they simply don't think to ask. And that's just in a classroom demonstration with one clearly defined objective (to obtain a hundred-dollar bill). It can be even harder to remember to ask the right questions during live conversation, when questions do so much more than just ask *for* something that we want. Conversational questions can pursue motives all over the conversational compass, from selfish requests *for things* to selfless expressions of interest. They can simultaneously invite information from a partner, while (at times, imperfectly) revealing the motives of the asker. They can be profoundly responsive, with each question reacting to a turn or series of turns that came before. They control the flow of our conversations, steering us toward the next turn and the next topic. They help

---

* The winner humbly returns the money once the exercise ends or after class (knowing that professors can't give them cash gifts). The real prize is winning the game.

us bide our time when we don't know what to talk about, and they can determine the quality of our connections and, ultimately, our relationships. With all these different functions, amid the coordination chaos of real-time conversation, there's so much room for error. It's no wonder we simply forget to ask them in the right moments, just like my students forget to ask for the hundred-dollar bill.

To make our way through the complexity, it's important to understand the different *types* of questions we have at our disposal. Understanding question types not only helps us notice their absence and remember to ask them, but also helps us remember to ask the right questions in the right moments. Ideally, we want a typology that takes into account the unfolding, flowing nature of conversation and that sorts questions by the role they play in the context of the conversation. Such a typology can help us understand how to be more responsive to our partners and more generative about where things will go next.

When my fellow researchers and I analyzed fifteen-minute get-to-know-you conversations among 398 strangers, we found that four dominant question types emerged: *introductory, mirror, topic-switching,* and *follow-up*. The graph on page 72 shows how often people asked each question type over the course of their conversations.

So how do we distinguish these question types? *Introductory questions* are what they sound like—"What's your name?" "How are you?" or "What's new?" They come near the beginning of our conversations and are the entry points to small talk. They help the speakers orient themselves to each other and the current moment. They're at the base of the topic pyramid. These common, habitual questions can be useful, but the goal, as with small talk, is to move past them quickly. As I'm about to.

Mirror questions reciprocate a question that's just been asked, as in "I'm good. How are you?" They're not always insincere, but they're often used more out of politeness norms than out of sincere curiosity—and it's hard to tell the difference when you're on the receiving end.

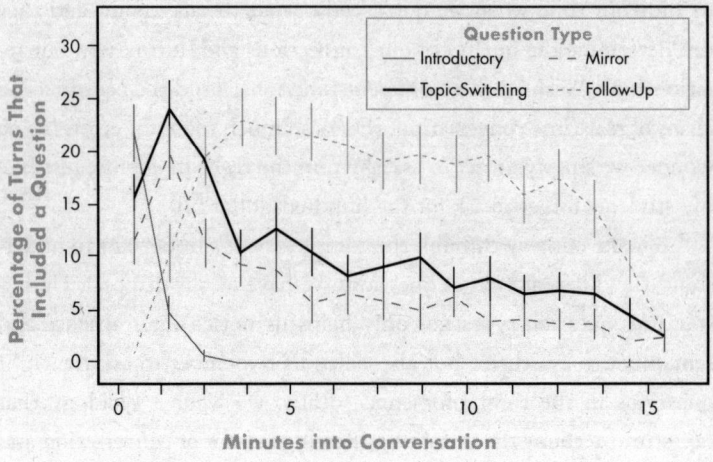

How people ask different types of questions over time

Like introductory questions, they tend to be dictated by the norms of conversation. You asked me something, so I feel like I should ask it in return. Since they don't necessarily reflect authentic curiosity and care, responders shouldn't linger on their answers—and askers may want to figure out a better way of reciprocating, rather than just repeating, verbatim, a question someone has already asked.

Research provides us with concrete reasons to move on quickly from introductory and mirror questions. Remember the finding that asking more questions increases likability? Before you ask rapid-fire introductory or mirror questions to boost your conversational "likes," hold on a second: it turns out this doesn't apply to *all* questions. The boost in likability *isn't* triggered by mirror questions or introductory questions. That might not mean much when it comes to introductory questions, which are necessary and important (just like small talk). But asking more mirror questions doesn't increase likability, either. All the more reason, then, to ask them rarely, in those moments when we truly want to hear the answers, not just out of politeness or convenience.

And yet mirror questions persist deeper into a conversation than we might expect. In our study tracking question types over time, we found that while introductory questions tapered off quickly, mirror questions, despite their limited benefits, decayed more slowly and continued to be asked at a moderate rate throughout the conversation. That's because mirror questions are easy. They keep the conversation flowing with little effort or creativity. When you aren't sure what to say next, mirror questions are low-hanging improvisational fruit. They're useful in a pinch. They're also okay when you really want to hear a partner's response, closer to the top of the topic pyramid ("What about *your* relationship with your mother?"). But you shouldn't rely on them too much, or for too long, or without adding some interesting new twist. We have other, better tools in our question-asking arsenal.

Ready for a new topic? *Topic-switching questions* initiate a noticeably new topic, and in doing so, they end the previous one. In the last chapter, we talked a lot about when and how to switch topics—asking topic-switching questions is the most common way people switch. Here's an example from one of my studies. A pair of strangers is getting to know each other. To start, the first speaker asks an introductory question: "How was your day?"

To which the partner responds, "Work was a little bit stressful, but otherwise it was good." It's a standard setup for a mirror question ("How was *your* day?"), but instead this speaker jags to a new topic: "Did you see that the new rover landed on Mars today for NASA?"

By not taking the bait for a mirror question and instead using a topic-switching question, this speaker climbs the topic pyramid rather than floundering at the base.

Coming up with topic-switching questions is an excellent skill to practice. Once we think of a topic, our brains can get really good at formulating questions to raise it. For example, *meatballs:* "Do you like meatballs?" "Do you ever cook meatballs?" "Does your family have a good meatball recipe?" "Have you ever tasted meatballs you didn't like?" "Is it weird that I've come up with so many questions about meatballs?" Whenever you're brainstorming topics—either before the conversation or in the middle of it—you can lean into questions as an easy way to frame a new topic to reinvigorate the conversation.

When my fellow researcher Mike Yeomans and I measured topic-switching questions over time, we found that people asked more toward the beginning of the conversation, then fewer and fewer as they settled into topics that seemed interesting. It makes sense. But it's also likely a missed opportunity—as we've seen, many people could stand to switch topics more frequently. Topic-switching questions are the most reliable way to do so.

The Incredible Inquirer

While topic-switching questions are our friends, *follow-up questions* are superheroes. Follow-up questions are the opposite of topic-switching questions: they keep us on the present subject, probing deeper on something a partner has previously said. They help us explore a topic once we're there together or have been there together in the past. While discussing meatballs, a partner might ask a follow-up question like "Did your grandmother have a lot of special recipes?" "Do you know how to cook all her recipes?" or "What's your favorite of your nonna's recipes?"

Follow-ups help us delve deeply into a topic, sometimes quite quickly. In our studies (the ones that showed that more questions led to higher liking), when we prompted people to ask more questions, the extra questions they came up with were mainly follow-up questions. This is key—it meant that the rewards of asking more questions were driven *almost entirely* by asking more follow-up questions. These questions offer affirmation: "I heard you say that you like meatballs. Can you tell me more?" They unlock deeper learning: "Why do you love meatballs? Whose do you love most? When was the last time you had really good ones?" They are inherently personal and validating—they make people feel heard because they show that they've actually *been heard*.

In an exercise for my class called Never-Ending Follow-Ups, I challenge my students to have a conversation in which they *only* ask follow-up questions. Whenever they decide to take on this challenge outside class, they're allowed to make statements, too, for the sake of smoothness, but they must always respond to what their partner has said with a follow-up question. The goal is to get them asking more follow-up questions, and for them to see how easy and powerful it is to do so.

In a sense, the exercise turns each student into a version of Terry Gross, engaging inquisitively with any partner of their choosing. Sometimes the conversation partner catches on, but many times they don't notice the game. When we're asked a question, our minds become busy figuring out how to answer—too busy to notice the conversational jujitsu our partner has pulled on us. My students, now jujitsu masters, are surprised by how easily they can achieve never-ending follow-ups, how much they learn about their partner in the process, and how fun and engaging it is—for everyone involved. As the experience makes plain, follow-up questions can be used consistently throughout a conversation. Doing it every time you talk is a bit extreme, but as my students see, it's not nearly as crazy as we might think.

Great conversationalists strike the right balance between topic-switching questions (to hop to new topic islands) and follow-up questions (to explore the islands once you're there). While in general we should aim to switch topics whenever the conversation starts to lag, that doesn't preclude using more follow-up questions as well—we can do *both*.

Ironically, asking mostly follow-up questions doesn't mean talkers stay on the same topic the whole time. Follow-up questions can lead a partner through an exciting drift, opening new worlds and topics, even as they directly spring from what the partner has previously said. This is another power of superhero follow-ups—they're also stealth topic-switchers. While one person is asking a follow-up question, the other is answering it—drifting logically from meatballs to Italy to family heritage to their long, fascinating lineage of scientists, doctors, opera singers, and outrageous heretics.

Learning from the Masters

Though almost any follow-up question can keep a conversation alive, it can be hard to know how to ask ones that will lead to the best treasure. Here we can take a cue from Terry Gross, who uses follow-up questions skillfully. If her subject isn't taking us in an interesting direction, she usually tries one or two follow-up questions. If still no treasure is revealed, she moves on, often going in a different direction to an unrelated topic. She is curious to strike gold, but appropriately impatient—always ready to abandon the search to jump to a new island.

We can learn more useful approaches from other master question askers like Oprah Winfrey, who often probes with follow-up questions about emotion, like "How did that make you *feel*?" Barbara Walters used follow-up questions to explore past-present-future scenarios, asking her interviewees to travel through time to revisit old decisions, reflect on their current feelings, or make predictions about the future. Walters interviewed Monica Lewinsky in 1999:

BW: If you had to do it all over again, would you have had the relationship with Bill Clinton?

ML: There are some days that I regret that the relationship ever started and there are some days that I just regret that I ever confided in Linda Tripp.

BW: Monica, are you still in love with Bill Clinton?

ML: No. . . . Sometimes I have warm feelings, sometimes I'm

proud of him still, and sometimes I hate his guts. And, um, he makes me sick.
BW: What will you tell your children, when you have them?
ML: Mommy made a big mistake.

Past-present-future follow-up questions help speakers examine decisions and feelings from multiple angles. They break people out of their natural myopia (focusing narrowly on the present), which can lead to fascinating new insights and possibilities. "Are you on vacation now? . . . How is it? . . . Where did you go for your last vacation? . . . And how about the next one?" Past-present-future framing help to pull out feelings of nostalgia, reflection, and hope in quick succession.

One reason conversational masters like Gross, Winfrey, and Walters seem so confident, even in extremely tense conversational environments and on difficult topics, is that they are follow-up masters. They know they don't need specific expertise—deep knowledge of a topic—to feel comfortable discussing it, because they can ask follow-up questions. It's the ultimate improv tool, *especially* when they don't understand or know what to say. Unlike topic forethought, follow-up questions can help us keep the conversation alive as we go along. When in doubt, if you don't know what to say, be a journalist like Oprah Winfrey, Barbara Walters, or Terry Gross—and just keep asking. The key to good conversation isn't knowing, but learning. It's about being interested (in your partner), not interesting.

What's with All the Questions?

In my classes, after learning about the apparently unlimited power of topic-switching and follow-up questions, someone always voices the same fear: "I worry about asking too many questions and being annoying."

Sometimes this observation comes from a student who isn't from the United States. When that happens, I acknowledge that different cultures may have different norms about how many questions you're expected to (or allowed to) ask, and who's expected to ask and answer them. Cultural nuance is always a crucial part of reading the room. The emerging science of conversation doesn't yet have thorough evidence about question-asking norms around the world.

But in the United States, we have some evidence to work with. When you realize that we should ask more questions—that we should ask more topic-switching questions and can ask follow-up questions indefinitely—it's only reasonable to wonder: Is there a tipping point when *many* questions become *too many*?

Let's go back to our study of speed daters. My fellow researchers and I looked at this very issue, trying to find the place where the next question was one too many. The answer was exciting: we didn't see one! People didn't ask enough questions to get there. In fact, those who asked ever more questions were *ever more likely* to get invitations for second dates. The same was true in our studies of platonic strangers getting to know each other: asking lots of questions improved the conversation and made the asker better liked. These findings suggest that in many everyday conversations, the "too much" tipping point can be quite high, possibly unreachable.

But like everything regarding conversation, the tipping point likely depends on context. On speed dates and during casual chats between new acquaintances, talkers have *so much to learn* about each other, and their goals—to be polite, to talk about themselves, to have fun, to learn about the other person, to not be awkward—often align. But some conversations, in theory, might have lower tipping points: they are more tense, competitive, conflicted, uncertain, or suspicious. In such contexts, asking questions can feel less like the asker is interested and more like they might be judgmental, adversarial, or exploitative—and that answering their questions may put the respondent at a strategic disadvantage.

For instance, imagine a girlfriend who is trying to use questions to get to the heart of her boyfriend's hesitation to go to a party together. She's trying to suss out the truth. Does he think he'll be too tired from work? Will there be someone there he doesn't like? A former flame? Is he embarrassed by her? She wants to understand so they can make a fair decision, but her questioning makes him more and more defensive. The conversation ends in frustration and anger, and they are no closer to a satisfactory plan.

Tense conversations with seemingly too many questions don't happen only at home. Think of a salesperson calling a potential customer. The salesperson wants to learn as much as possible, then leverage that information to persuade the potential client to talk again or to agree to buy—they want to keep the lead alive as long as it's still promising. The customer, meanwhile, wants useful products and services that fulfill their needs—at a price they can afford. These two sets of goals are partially aligned (the salesperson wants to sell, the customer wants to buy) and partially in conflict (the salesperson wants a high price, the customer wants a low price). It's a negotiation—a delicate dance of information exchange and relationship development. Disclosing too much information could damage each party's position. Once a salesperson knows the maximum a customer can afford, for example, they're likely to price their goods right at this amount, if it's in the "zone of possible agreement."

For these reasons (and many others), sales is a specific context that can be competitive, unpleasant, adversarial, and annoying. So it would make sense that a salesperson (or a girlfriend trying to persuade her boyfriend to accompany her to a party he doesn't want to attend) should be more careful about asking too many questions than someone on a speed date.

I work with a company called Gong that seeks to untangle problems like this—how many questions should people ask?—by recording and analyzing huge numbers of real sales calls. Gong's product allows companies to understand which sales strategies are working and which ones aren't, ultimately allowing salespeople to improve their conversations. But for our purposes, what's exciting about working with Gong is that we can see how question-asking plays out in a tricky context across lots of people. We can learn to emulate the salespeople and customers who achieve the best outcomes.

One of the many behaviors Gong analyzes from sales calls is question-asking. They, too, find that more questions are good—those who ask more questions learn more useful information during discovery calls, convince their customers to talk to them again, and ultimately convert more calls into sales. But unlike our studies of speed dates and negotiations, Gong's data do show a tipping point. In one sample, for example, asking more than twenty questions every ten minutes was worse than asking fifteen to twenty questions. When customers are bombarded by questions, they likely feel overwhelmed, annoyed, and perhaps defensive.

Let's hold on a minute, though. When we stop to think about it, twenty questions every ten minutes is *a lot of questions*. That's two questions per minute, on average, for the duration of the call—a relentless pace. It's hard to imagine asking that many questions in the kinds of casual or intimate conversations we've been focusing on.

Yet even at this relentless pace, asking more than twenty questions every ten minutes was *still* better than asking fewer than ten. In sales calls or other charged conversations, while it may be possible to overshoot and ask too many questions, just as in most conversations, the real danger is asking too few.[*]

---

[*] Gong also finds that the most effective salespeople tend to spread their questions evenly throughout their calls with a steady stream of follow-ups, while average salespeople tend to front-load their questions at the beginning of the call.

Are There Any Bad Questions?

You've probably heard teachers say, "There are no bad questions!" to encourage their students to ask more. While we'd love to think that there are no bad questions, sadly, this isn't the case, and we need to be aware of it. Actually, it's not so much that there are bad questions, but that there are bad *patterns of questioning*—bad because they're counterproductive when it comes to achieving our conversational goals (unless you want to make your partner think you're a jerk or convince them to break up with you). For instance, a participant in one of our studies asked his conversation partner, "Have you won any contests lately?"

"No," his partner responded.

And then the asker swiftly said, "I won the Pick a Present contest that will happen in December!"

Unsurprisingly, the asker also won the award for worst question in this conversation. (I made that up just now and bestowed it myself; his trophy is en route.) He asked a question—an oddly specific one—for the sole and obvious purpose of answering it himself, so that he could brag about his own good luck. This is a nasty but common habit that my fellow researchers and I call *boomerasking,* named after the outgoing-and-returning, just-can't-stop-itself, self-centered arc of a boomerang. Motivated by the dual desires to disclose self-centered information and to not seem selfish, boomeraskers pose a question like "How was your weekend?" only to answer it themselves, soon after their partner's response, "Cool, cool. Well, I spent the weekend skydiving with Harry Styles."

In our conversation datasets, boomerasks are common—and problematic. They convey a lack of awareness, motivation, and attention to the partner's response. The boomerasker doesn't listen enough to their partner, or insufficiently acknowledges their partner's disclosure, or both. They take everything a question is supposed to do, and they achieve the opposite. Boomeraskers aren't always mean-spirited, but it's the impact, not the intention, that ultimately matters. Even if the initial question was born of sincere curiosity, the answerer, whose answer is swiftly ignored, sees the sequence as uncurious and uncaring—the boomerasker seems insincere. Our studies of boomerasks show that whether we want to brag, complain, or simply share information, we're better off just saying it explicitly, rather than prefacing it with a question under the pretense of sincere interest.

Boomerasking is just one example of how question-asking can go awry by conveying the impression of less-than-admirable motives. Questions that spring more from our own insecurities ("Do you like the dinner I made?") or from a subconscious desire to hurt someone ("Dave, do you like the dinner Clara made?") don't go as well as questions that spring from curiosity and care ("Clara, how did you make this meal?").

It's not that these questions always reveal our deepest, darkest bad intentions, though they definitely can. Rather, they make us *seem* like we have bad intentions, even if we don't. Questions have a way of revealing our conversational goals, even when we don't mean to reveal them or are not even fully cognizant of them. Questions that spring from unvirtuous, unclear, or conflicted motives risk conveying a negative impression. If you find that a question or series of questions has seemed to reveal your selfish or unsavory motives, it may be helpful to check in with the conversational compass. Ask yourself: What was I trying to achieve by asking that question? Is that really the kind of person I want to be?

When we ask sensitive questions, we usually fear offending our recipient, but when we ask bad questions we look bad (and can spoil the mood). And while sensitive questions that are underpinned by curiosity and care tend to be received more favorably than we anticipate, bad questions—underpinned by self-serving or insincere motives—don't land well. The difference is in the recipient's perception of positive or negative intent: *Does the asker seem to be asking because they're interested and care about my perspective, or because they're required to? Because they want to embarrass me? Or because they might want to use my answer for their strategic advantage?*

Along with boomerasking, a couple of other kinds of question-asking are likely to be received negatively. Gotcha questions—like "You said you know all the state capitals, so what's the capital of Nebraska?" or "You said you care about gay rights, so when's the last time you supported an LGBTQ+ bill?"—are acutely threatening. They make respondents feel like the asker wants to prove them wrong or make them look and feel incompetent. Such questions may have a useful place in political sparring, but in real, private conversations, they can make answerers feel *pushed* or *tested*—a quiz that they might flunk—and cause them to infer malicious intent.

Similarly, asking the same question more than once, maybe in slightly different ways, but without altering what you're asking *for*—can likewise make your partner feel defensive. Repeated questions are different from gentle probes for information via follow-up questions. They feel more like the asker is a dog with a bone and won't let it go.

Of course, repeated questions, gotcha questions, and boomerasks aren't *always* bad. What matters is how these patterns of questioning are perceived by the listener. Some people might be fine with being quizzed about state capitals, or they might feel grateful to be asked the same question several times to make them think about their answer more deeply. Maybe they're happy to hear their partner brag about Harry Styles. While it's probably best to avoid questions that have a chance of being perceived as insincere or aggressive, in the right context—like with close friends in warm, trusting relationships—even "bad" questions are usually fine (and with the right context, they might be used to great comedic effect).

Open and Shut (By Which I Mean *Closed*)

Do you need a little break? Are you close to the kitchen? During your break, if you were forced to choose, would you rather grab a cold drink or a cookie? In written text, these questions might seem refreshing or thought-provoking. But in conversation, you should use questions like these—which offer a predefined set of responses (closed questions) rather than unlimited room to speak freely (open-ended questions)—with caution. When you ask a friend, "Have you been to that new restaurant yet?" this closed question signals that the response should be short—yes or no will suffice. Closed questions can be a problem because they give the impression that you aren't really curious about what your partner has to say, that you want to keep things short and transactional. It's possible to frame this question in an open-ended way: "What do you think of the new Mexican restaurant?" Asking an open-ended question gives your partner more latitude to respond at any length they want, get creative, or steer the topic in a different direction.

Talkers are good at taking these cues: on average, people respond to open-ended questions with twice as many words as they use to respond to closed questions. Of course, closed questions aren't all bad. They're efficient—they can be helpful when a speaker needs to quickly figure out what others know, like asking "Have you seen the most recent episode of *The Handmaid's Tale*?" before launching into a spoiler-filled diatribe. They help us extract very specific information. But if you want to make the conversation fun and smooth, and to learn things about your partner—to foster a more open and interactive dialogue overall—then open-ended questions are better.

In practice, we sometimes intend to ask an open-ended question, but because we feel impatient, excited, or expectant, we immediately provide a set of options: "Why are you a writer? Did you grow up wanting to be a writer, or did you fall into it?" By narrowing the response set, we've turned the open question into a closed question, with two possible answers. These are what the conversation analyst Anita Pomerantz has called "candidate answers." Though they can be helpful when interacting with children, giving candidate answers to adults is annoying. We should trust our own questions and let people answer them, without leading (or stifling) the witness.

Together with colleagues, I've studied open-ended and closed questions during negotiations, a conversational context that is defined by disagreement and that, like sales calls, can make people hesitant to ask and answer questions. One study involved a role-play between the executive editor and the advertising manager of a midsize newspaper—a simulation that mirrors many typical budgeting negotiations. While the executive editor was primarily concerned with improving the paper's quality, the advertising manager wanted to increase advertising revenue.

We tracked hundreds of people's behavior in this negotiation role-play. People who asked more open-ended questions ("For you, what is the most important thing?" "Where would you like to start?" "How do you see it?" "What do you mean?" "What's another solution?" "How would that work?" "Can you tell me more?") learned more information, were better liked, and were more likely to discover opportunities to make productive trades or concessions. In a follow-up experiment, we randomly instructed negotiators to ask more versus fewer open-ended questions. We found that asking more open-ended questions actually *causes* those positive outcomes. Ideally, conversationalists across a wide array of contexts—both cooperative and conflicted—are best served by asking more open-ended, follow-up, and topic-switching questions.

Why Do You Ask?

If asking follow-ups is a superpower, it's our responsibility to use that power for good. Asking follow-ups—along with asking more open-ended and topic-switching questions—can sometimes seem to risk turning question-asking into a party trick or a cynical ploy for personal gain. What if you don't *want* to ask a follow-up question about your talk partner's weekend? Should you fake it for the sake of a successful conversation? The idea of gaming the conversation, especially for personal gain, makes us very uncomfortable—it feels inauthentic, manipulative, or deceptive. And most of us worry (a little or a lot) about both being and, ironically, seeming authentic and honest.

But this line of thinking is another tentacle protruding from the Myth of Naturalness (see Chapter 2). We come to believe that good conversationalists ask only fascinating questions that they're dying to hear the answers to. In truth, good conversationalists often ask questions to keep the conversation from dying.

The remarkable thing about keeping a conversation alive is that fleeting moments of insincerity can lead to breakthroughs of authenticity—in fact, it's a good way to get there. For people with good intentions overall, asking *anything*—however basic or banal—to keep a conversation alive and to avoid the dreaded ZQ is fine. Actually, it's *great*. Asking a polite question without profound interest behind it needn't feel like a meaningless party trick or a cynical manipulation. Asking itself is a form of caring—an acknowledgment of your partner's unique mind and your continued willingness to engage with whatever they might say back. It shows your commitment to uncovering treasure together.

And here's the best part: the conversation *will* keep going. No matter what you ask, your partner will answer your question in one way or another, and you can build on their answer. Asking "How was your weekend?"—even if you don't care that much and you know it's a bland question—might remind you that your partner belongs to a rock climbing gym or book club, something you really admire. Then you can ask about their experience, who they climb with, what they like or dislike about it, and their advice about getting into it. You never know when one unassuming question will trigger mutual curiosity or unearth valuable treasure. It's like starting with small talk and getting to deeper things—you have to keep the coordination game alive if you want to win its richest rewards.

> THREE KEY TAKEAWAYS FROM CHAPTER 3
>
> ## A Is for Asking
>
> - Aim to **ask more questions**. Asking even insincere questions is a form of caring, and asking too many questions is rare.
> - Use caution with **boomerasking, gotcha questions,** and **repeated questions.**
> - Do **ask topic-switching questions** to change topics and **follow-up questions** to learn more and keep the conversation alive.

CHAPTER 4

L Is for Levity

HAVE YOU EVER BEEN TO a restaurant, and at some point you realize the couple next to you is on a first date? The searching questions, the lingering smiles, the subtle (and sometimes blatant) flirtations. The way people put themselves forward when they're trying to impress someone. Oooh, those awkward moments when a joke doesn't land or a topic isn't clicking. Your ears are perked, eyes stealing sidelong glances as the show unfolds. You're desperate to hear the next line. You, and perhaps the person with you, are amused by the intermittent eavesdropping, even as you pretend to be completely oblivious. *Is he still talking about crypto? Can't he see the look on her face? They're so nervous! They're trying so hard!*

Now imagine sitting next to a thousand first dates. That is part of my job—one of my favorite parts. Scientists have studied romantic dates, interpersonal attraction, sexual selection, and marriage for many years (decades). But this research has missed an important piece of the puzzle: how people talk to each other. The new science of conversation takes this research a step further by recording and analyzing the transcripts of speed dates, so we can go beyond *who* is appealing to uncover *why* they are appealing. With their high stakes, speed dates are a wonderful way to see in action much of what we've learned so far.

We've already heard about some speed dates, like our star-crossed, weather-talking nonlovers Josie and Hugo in Chapter 2. But what do good dates sound like? How does conversation influence interpersonal attraction? Do people know when their partner is flirting with them?

The largest dataset of dating conversations was compiled by a research team led by Stanford scientists Daniel McFarland, Dan Jurafsky, and Rajesh Ranganath. In a great service to both science and the world, they conducted three speed dating sessions—eleven hundred dates—in a California restaurant in 2005, the year when "Hollaback Girl" by Gwen Stefani was the most popular song on the radio, Brad Pitt and Jennifer Aniston divorced (then Brad quickly got together with Angelina Jolie), and President George W. Bush was sworn into office for his second term. The daters, heterosexual graduate students, wore shoulder sashes with microphones, volunteering in exchange for the promise of email addresses from any daters who liked them back. The dates were in an open restaurant, with lots of background noise—just like most dinner dates—except each of these dates lasted only four minutes. It's a dream scenario for anyone who's ever *wished* their date was only four minutes long, but it's also the perfect, albeit compressed, way to study first dates.

The students were strangers to each other—these folks were going in totally cold. They weren't connected through mutual friends. No online profiles were reviewed beforehand. It was before dating apps became popular with the mainstream. They'd signed up to do this to meet new people, with no guarantee about finding commonalities, only the hope that they might.

Two daters sit across from each other at one of many small tables in the noisy restaurant—Tom and Cassie. They've already spent some time moving around the room and talking to others, so they know the drill. As soon as the timer goes off, they'll have four minutes to get to know each other. Then they must move on.

Before the timer even starts, the two get in some quick words, joking over how inept they feel at this speed dating thing.

"I don't think I'm really doing very good at any of this," Cassie offers.

"Yeah, I messed up," Tom says. "I totally screwed up the first five or six."

They laugh, and she asks if he has enough water.

"Oh, I never had any water to begin with," he replies. "I don't know how everyone else has water except me." They laugh again.

Still waiting for the timer to start, the two continue. "It's like a rotating merry-go-round," she says. "I feel like I'm just reciting lines from a script."

Tom agrees. "So you're not going to become a professional speed dater?" he asks.

They laugh again.

The buzzer goes off—time to get to know each other (as if that hasn't already started).

"We're allowed to talk now," Tom announces. But instead of getting down to business, he revisits their shared plight. It's so awkward, he says, to speak into a recorder. He's burning precious time. "I don't think they're getting, like, good data," he laments.

They laugh again. So far, he's kicked off the conversation with four innocuous jokes. Perhaps encouraged by the laughs, he offers a new topic: he's studying computer science.

"Really?" Cassie replies. She wouldn't have taken him for a computer science guy.

"What would you have *thought* I was studying?" he asks.

She doesn't know. Tom reveals he studied English as an undergrad. So did she, it turns out—she even taught English for a while.

Cassie returns to his question. "What would I have *thought* you were studying?" she says, mulling. "Yeah, like English."

"Yeah, I don't like computers. I hate them," he says.

"Me too."

"They suck."

They laugh again. This exchange makes him sound both romantic for studying English and practical for pursuing something more lucrative—all couched in a self-deprecating joke that he hates computers, perhaps showing that his heart is more aligned with hers than it might have first appeared.

They trade a few notes—she's studying law, and he's in computers because he didn't want to get PhD on the technicalities of *Hamlet*. She misses teaching.

Then she asks where he's from.

"Florida," Tom replies.

"I'm from the South too," Cassie says. "Although Florida is not really the South, but—"

"No," he agrees. "Florida is just a weird place." He's using self-deprecation to support her position. And he wants to know where she's from. "Are you from Texas?" he asks.

"Oh, no. I hate Texas."

"Louisiana?"

"No."

"You said from the South."

Yes, she says. This is a fun game.

"Georgia?" Tom asks.

"Well, you're getting closer," she replies. "But *Texas*?"

He mumbles something.

"I *despise* Texas," Cassie says.

"I thought Austin was a pretty cool place," he offers.

"You know what, I have heard that."

Well, Florida is probably worse than Texas, he tells her. In fact, Florida is awful.

"I'm from—" she starts.

But Tom's not done guessing. "What, Tennessee?"

"No."

"South Carolina?"

"No."

They keep going for a bit, until finally, he gives up.

"So where *are* you from?"

"North Carolina," she says. "The better Carolina."

"Well, that's true," Tom agrees.

They compare some thoughts about where they grew up, the complexities of the South, and then: Four minutes. Time's up. The date was a success. In a private survey, they both tell the researcher that they want to see each other again. They're given each other's email addresses at the end of the event. They also rate each other highly on metrics of date success, like how much they clicked.

There's a risk to overanalyzing this conversation based on the dialogue alone. For all we know, they both found each other ridiculously hot and could have exchanged *um*s and *oh*s for four minutes and still wanted to see each other again. But their conversation, studied more closely and compared against all other speed dates, yields clues to their chemistry. Though this is a high-wire conversation with no previous context between the two, they bonded quickly over the shared experience of the speed dating environment. She humbly admits she's "not doing very good at any of this," and he agrees that he's "messed up the first five or six." They've found shared reality, with little to go on. Affirmed and relieved, they laugh. Then they joke about not having any water. They laugh again. "It's so awkward." They laugh again. Is it possible that their admission of awkwardness is making this date *less* awkward?

Then, as they pivot to sharing their undergraduate majors and where they're from, they find more substantive common ground, assuredly climbing the topic pyramid. Though neither is *currently* studying English or hails from the same state, they both prefer English over computers, and they both grew up in the South. If we mapped their topics, the conversation opened with a string of minor, innocuous jokes, then predictable but successful common-ground-finding, and then the cherry on top: it closes with a game, invented on the spot. As they revise and deepen their common ground, they're playful—they gamify. He wonders about her *expectation* of his major, comparing it with his actual major. She makes him guess where she's from, holding on to the truth like a carnival prize. They're playing, and in doing so, they're winning.

The Power of Levity

One of the biggest challenges of a first date: Don't let the conversation die. If the conversation dies, the date dies. Even a two-second pause during a four-minute date can make you feel like it's going to be a long night (or at least a very long four minutes). The prospect of an awkward silence or the vibe taking a nosedive is terrifying, and the experience of it is *excruciating*—embarrassing, alarming, demoralizing.

Regular life may not always be as high stakes as speed dating, but conversations often feel that way. Maybe you raised a juicy topic, but then things got stuck on mundane details. Maybe your team is trying to address something serious and important, but the conversation is growing so glum that no one is brainstorming solutions. Your cocktail chitchat is starting to sputter, and everyone seems to be scanning the room for an out. And you know instinctively that all of you desperately need some spice. A friend is feeling sad about a breakup and needs someone to lift her spirits to help her move on. The eulogies at a funeral have gotten increasingly serious, and you're pretty sure the deceased would want their loved ones to be happy. Your work buddies are stuck on idle shoptalk. New parents can't seem to escape an endless discussion of naps and feeding schedules. How can we break out of these depressing everyday doldrums?

To make this shift, it helps to think more deeply about our feelings. And when it comes to thinking deeply about feelings, scholars find it helpful to take the complex constellation of human emotions and plot them on two simple dimensions: arousal (high versus low physical signals of energy, like heart rate) and valence (pleasure versus displeasure). All emotions—whether experienced during conversation or outside it—can be plotted on this chart of emotions. Here's where some common feelings fall:

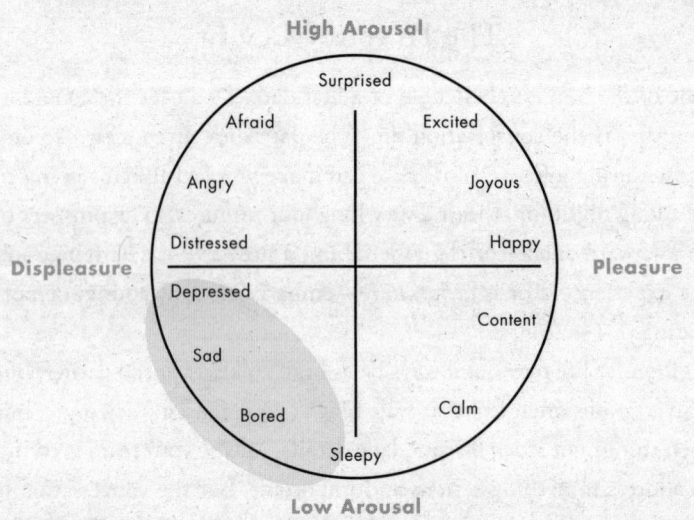

When we talk about the power of levity, we want to focus on two quadrants: the lower left and the upper right.

Let's zero in on the lower-left quadrant. We're dispirited. We're sad. We're bored. We're decidedly *not* energized. *Yawn.* What does it mean when our conversations turn this way? For starters, we disengage, even if we don't mean to or want to. Our minds wander. We make less eye contact. Our words get less positive and become more neutral or even negative (fewer *awesome, cool, wow* and more *sucks, ugh, the worst*). We speak less overall, which means we interrupt less but also have longer pauses between turns and fewer moments of authentic, well-timed laughter.

In the upper-right quadrant, we're the opposite. Here feelings of engagement and enjoyment flourish. We lean forward and match each other's gaze. We give the bubbling reactions of back-channel feedback, like *yeah, uh-huh,* and *oh!* We respond more quickly. We even interrupt more often, but it's because we're excited to find out what will happen next.

Disengagement poses a serious (relentless) threat to *all* conversations, while engagement channels the power of mutual attention and makes us feel connected and alive. This is where our daters were aiming. This is where we're all aiming a lot of the time, in all different kinds of conversations.

This is where the third maxim of successful conversation comes in: Levity. Great conversationalists strive for the upper-right quadrant, and levity is one of the best tools we have to get there. Like the yeast that turns a woefully dense puck of bread dough into crusty, fluffy heaven, or the helium that keeps a party balloon bobbing cheerfully in the breeze, or the pop of bubbly champagne that sparks excitement at a dinner party, levity is any conversational move—playful, funny, unexpected, warm—that infuses positive energy. It's the buoyancy and the fizz that keep our minds engaged and our senses primed, moving us from the lower-left to the upper-right quadrant.

Levity moves us beyond Topics and Asking because it focuses even more squarely on pursuing relational goals. We don't want just to exchange useful information, we want to enjoy ourselves while we do it. We don't want just to play the coordination game, we want the game to be fun. Our professional matchmaker Rachel Greenwald, an expert in engaging conversation, likes to talk about a "mood elevator" that we ride up and down as we talk. When the elevator moves to lower floors, we need to press buttons to go back up—and because playfulness and fun are contagious, we're able to bring our partners with us. That's levity.

The Pursuit of Levity, in Pursuit of Other Things

The powerful thing about levity is that it doesn't just keep our conversations alive. It also helps us pursue many other conversational goals: effective brainstorming, giving and receiving advice, persuading people to agree with our views, asking and answering hard questions, signaling and earning status, conveying information in a memorable way.

This is because the enjoyment that levity brings—that fizz, that sparkle—makes us behave (and think) differently. As the economist Andrew Oswald and his colleagues have shown, when you make randomly selected people feel happier, they become 12 percent more productive—they answer more exam questions, they are no less careful, and they get more answers right. Similarly, the organizational psychologist Teresa Amabile has shown that feeling happy provokes greater creativity, leading people to generate more ideas—with a higher proportion that are novel and actionable. In my own work, I've found that when people feel excited (instead of anxious or bored), they give longer, more persuasive public speeches and get more difficult math questions correct. They even sing better—louder, more rhythmically, and more on pitch.

"There is no better trigger for thinking than laughter," wrote the early twentieth-century German essayist Walter Benjamin, and that's not just because levity relieves anxiety or tedium. When we're happy and engaged, our mindset changes. We widen the scope of our attention. We consider more things we *could* do, which improves our creativity and the decisions we actually make. Our bodies also feel it. One good laugh can relieve physical tension and relax our muscles for up to forty-five minutes and lower our blood pressure significantly. Our recovery from stress, our immune function, our sensation of pain, our resistance to illness, even the length of our lives—all these have been shown to improve with frequent positive feelings.

Levity does more than just buoy us and unlock our best selves. It reflects, reveals, and reinforces authentic feelings of *psychological safety* and *trust*. Psychological safety—the belief that you won't be punished or humiliated for speaking up with ideas, question, concerns, or mistakes in a group (see Chapter 6)—is rooted in interpersonal enjoyment. We feel safe when we, and everyone around us, brings a spirit of play. Psychological safety (and trust in groups of two people) is a positive feeling tethered to the belief that it's okay to say or do things, even if they could make you vulnerable to negative judgment or exploitation by others. When we infuse levity in a conversation, we communicate this safety, allowing both ourselves and our partners to reveal more, share more, and co-create more. It's a virtuous cycle: playfulness makes us feel confident that it's safe to play. And that feeling of safety helps us pursue all our goals—in every quadrant of the conversational compass.

The behavioral scientist Leslie John explored this idea by manipulating the fun factor in people's environments. She gave people chocolate chip cookies, played music, or used a silly font—or not. These cues of levity, she found, were incredibly influential in how much people disclosed about themselves—a signal of psychological safety. When subjects were asked to complete a survey on a whimsical, colorful website, for example, they were 1.9 times more likely to reveal personal information—including their Social Security numbers and sexual history—compared to those whose survey interface looked serious and official. And disclosure (as we saw in Chapters 2 and 3) brings us closer, triggers mutual disclosure, and helps us climb the topic pyramid together.

Where's the Funny?

Despite the abundant benefits of levity, somehow most of us forget or hesitate to use this tool. Just as we don't ask questions as often as we could, we don't use levity enough. At age twenty-three, 84 percent of people report smiling and laughing a lot the day before. By age fifty, that percentage drops to 68 percent. It bottoms out at 61 percent around age eighty. Then we make a tiny rebound in very old age—because the aged are happier, or because happier people live past the age of eighty (or both).

As we age, we get less practiced at playfulness, perhaps because it's harder to practice. It's difficult to monitor which emotional quadrant we are in when our brains are busy doing so many other things, like listening, planning what to say next, and coping with the harsh realities of adult life. The noise! The traffic! Toilet paper! Global warming! Life insurance! Not to mention that a lot of adult contexts (read: workplaces) seem to disallow levity.

So levity plays an increasingly shrinking role in our lives as we get older, but people at every age underestimate its benefits, which causes us to underdeploy it in our conversations. When we feel focused, bored, and serious, we struggle to muster the cognitive energy to break out of the doldrums. And even when we want to lift the mood, we worry that others will see our attempts at levity as frivolous, incompetent, or disrespectful. In a discussion of serious topics in a serious context, jokesters might not be taken seriously, we think. Even people who think they use plenty of levity (or worry they use it too much) probably aren't using it as much as they think.

The thing about levity is, well, it's fun. It doesn't just make our conversations come alive, it makes *us* come alive. In the rapturous date between Tom (Florida guy) and Cassie (North Carolina girl), we can see reciprocal loops of happiness, disclosure, acceptance, and trust unfold before our eyes. Their transcript leaves a trail of personal breadcrumbs that's joyful to follow. Their behavior aligns with psychologist

Barbara Fredrickson's "broaden and build" theory, which asserts that positive emotions evolved as psychological adaptations that increased human ancestors' odds of survival and reproduction. While negative emotions narrow people's focus toward specific, urgent actions that are life-preserving (fight or flight), positive emotions widen the array of thoughts and actions called forth (to play, experiment, and explore), inspiring more ideas, more flexibility, more interpersonal attraction, and longer life. Improv comedians live by a "Yes, and" mantra—a phrase that reflects a mindset to always broaden and build on what their fellow performers say and do. "No, but" and even "Yes, but" can shut down conversations, fast. These principles—broaden and build, "Yes, and"—underpin our daters' flirty games. And they are exactly the things we need for good conversation.

So You Want to Be a Comedian

It's halfway through the semester in my TALK class. The students have practiced preparing and managing topics, asking and answering hard questions, and avoiding the dreaded boomerask. Today is the first day of a new section: Levity. Given my decidedly unserious teaching style, they appear in class excited for a humor workshop or some other shenanigans I've surely planned. They, like everyone, would like to be funnier. I know they're hoping for some secret humor formulas.

I open the class, arms wide. "Good morning!"

"Good morning." They smile back at me.

"I need to say something important up front: I can't make you funny."

They laugh. Surely I must be joking. Right?

Unfortunately, I'm not joking. It's really hard to give any kind of specific lessons on how to be funnier. The issue is that good humor is *acutely* context-sensitive. How many jokes would you be equally comfortable cracking in front of your parents, your old high school friends, your colleagues, your doctor, and your new flame? In a small intimate group over dinner, in front of a boardroom, in a public conversation with a bunch of strangers? Would any of the jokes you told last week, or last year, be funny today?

The context-sensitive nature of humor is what gives it power, sparkle, and electricity. In fact, the sociologist Erving Goffman singled out humor that is context-sensitive and thus unrepeatable ("you had to be there") as a bespoke gift to the people present. "Since the witticism will never again be as telling, a sacrifice has been offered up to the conversation . . . by an act that shows how thoroughly the actor is alive to the interaction." How noble! But I can't give my students universal advice about how to become alive to each unique moment in each unique interaction. This, I tell the students in class, would go against the nature of humor.

Humor's highly context-dependent nature makes landing jokes difficult and daunting. Nobody wants to hurt people—or get themselves fired. Nobody wants to be the one to make a joke that isn't funny. Again, this is partly why we underuse levity in our conversations.

But here's the good news: people's anxiety about conversational humor is overblown. My research with behavioral scientists Maurice Schweitzer and Brad Bitterly found that people *over*estimate how often and egregiously humor goes poorly—we conjure comedians who got canceled and poignant conversational moments in which humor went awry. And people underestimate how often attempts at humor go well—leavening the mood, drawing people closer, boosting perceptions of competence and status.

In my own research, I found that managers who were randomly assigned to make one joke in one conversation were over 9 percent more likely to be voted into leadership positions by their teammates. And bosses with *any* sense of humor whatsoever have been shown to be 27 percent more motivating and admired than bosses with none at all, and their employees 15 percent more engaged. We forget how good even minor humor can feel. My point to you is this: the incremental risk that one-off jokes will flop is much smaller than the aggregated risk of a lifetime of boring, disengaged conversation that leaves you (and those around you) miserable.

But our worries aren't totally irrational. The trick with humor is getting it right. And as we know, success depends on what we're aiming for—our goals. In the case of humor, it may be more beneficial to aim for *finding the fun* rather than being funny.

Take It Easy

I can't necessarily teach people to be funnier, but I can show them what *kinds* of humor work in pursuit of important goals like enjoyment and connection. In class, I tell my students that we'll work on learning helpful frameworks for levity—we'll come to understand the mindsets that funny people inhabit. When I say this, the hopefulness returns to their faces.

Psychologists Peter McGraw and Caleb Warren suggest a logical framework—the Benign Violation Theory of Humor—to help us understand the risks (and sweet spot) of humor. Their theory suggests that people find things funny when they are neither too benign (boring, mundane) nor too violating (scary, sensitive, raw, aggressive, inappropriate), but somewhere in between. For example, simply walking down the stairs would be benign; falling down the stairs and breaking an arm would be a violation; and pretending to fall down the stairs, limbs akimbo, and landing gently, uninjured, in a silly dance pose would be a benign violation.

Finding the benign violation sweet spot can be quite difficult in conversation because it unfolds so quickly, and because the sweet spot is a moving target. It's further complicated by the fact that we all have different humor styles. So a remark that might seem like a benign violation coming from sweet-most-of-the-time Olivia could seem completely mundane coming from sarcastic-most-of-the-time Jimmy. What might seem funny to slapstick-loving Sam might seem completely ludicrous to political-satire-loving Jane.

Research on humor styles suggests that we all have tendencies and levels of comfort along several dimensions of humor, starting with how we position ourselves. Some people like to feign arrogance for comedic effect and use humor to signal their greatness ("Gosh, I'm good at this"), while others use self-deprecating humor brilliantly to seem humble or disarming ("This babydoll dress makes me look like an actual toddler"). Likewise, some people are outspoken and gregarious, lighting up the room like the sun ("What's up, party people?!"), while others are more understated—adding barely noticeable glimmers here and there. You can probably think of examples of each type in your life.

But what's really key about conversational humor isn't so much its style as its intended goals when it comes to our partners. Whether our style is mock-arrogance or mock-humility, over-the-top or sly-and-subtle, we can use humor to build others up or to pull people down. This distinction is tightly linked to thinking about humor, broadly, as *affiliative* versus *aggressive*. Affiliative humor is meant to be funny to everyone, to bring people together (rather than tear them apart or fragment them into categories). Aggressive humor, by contrast, involves put-downs or insults targeted toward one or a few individuals. Affiliative humor increases psychological safety (an antecedent to, well, everything good), while aggressive humor decreases psychological safety. The advice here is clear: when in doubt, be gentle.

The Joke's on Me

The former U.S. congressman Ric Keller locates the real start of his political career with a joke he made at one of his campaign stops at age thirty-four. He was last on the docket of a long list of speakers.

After sitting through all the other speakers, he opened his speech with an impromptu joke: "I feel like Elizabeth Taylor's seventh husband on his wedding night. Technically, I know what I'm supposed to do. But at this point, I don't know how to make it interesting."

It's a cheesy joke, for sure, but that's the beauty of it: he got a huge laugh from the crowd, who was hungry for a fresh, more grounded voice after hours of hearing serious political pundits. I'm not saying that one joke put him over the top, but Ric did go on to win the election and serve eight years in Congress.

Herein lies the power of one of the key forms of affiliative humor: self-deprecation. Most of us think poorly of ourselves, at least sometimes. We shouldn't let all that self-loathing go to waste! Self-deprecation works best when leaders and high-status group members share something they've overcome or criticisms they've received earlier in their life. In Ric's case, he was new to the political scene, which made his self-deprecation particularly brave. But he learned to use self-deprecation more and more throughout his years in office—embracing his vulnerability even as he became less and less vulnerable.

It's important to note that Ric didn't tease the other speakers who came before him, or the audience. He could have poked fun at how boring or similar the other speakers were, the way comedians veil insults with humor to roast celebrities, and each other, mercilessly. But here's the catch: other-defeating humor is actually the sniper of levity. It has the capacity to take us in the wrong direction on the chart of emotions, killing the mood rather than lightening it. That's why it's generally best to avoid it—especially "punching down" by making fun of anyone with lower status.

Sure, it can be tempting to summon our inner Don Rickles, the legendary insult comic said to be the only person who could make fun of Frank Sinatra to his face. One of my favorite *New Yorker* cartoons shows a father taking a knee and instructing his son: "If you can't say something nice, say something clever but devastating." And sure, perhaps on the rarest of occasions, we can make fun of someone we love in such a way that they're in on the joke, and ultimately it brings us closer.

But that's a tiny needle to thread! If you're going for levity and are considering a choice between gentle-but-cheesy and clever-but-devastating, the better bet is the former. Why? When it comes to levity, it's less costly to flop for being too benign than for being too aggressive. While the risks of being too benign are that people don't laugh and everyone moves on, the risks of being too aggressive are hurting people's feelings—or sense of self. Everyone likes to laugh, but no one likes to be laughed *at*. The pain from even fleeting moments of aggressive humor can take a long time to heal, and sometimes never does.

Make 'Em Laugh

I tell my students that I can't teach them to be funny, but funny people bring certain mindsets to the world—ways of seeing the potential for levity everywhere. I can't promise that these principles will turn out the next generation of *SNL* writers or land you a Netflix comedy special, but these mindsets might help you feel more prepared to try a joke you otherwise wouldn't in your next conversation.

THE SEINFELD STRATEGY

How can we add oomph to the benign? Here paralinguistic cues—tone of voice, pausing, timing, cadence—can be quite powerful. Paralanguage is often what people mean when they say it's all about delivery or timing. For example, Jerry Seinfeld, as a fictional character on the 1990s show *Seinfeld,* made humor magic out of everyday observations. Using what Peter McGraw calls the "Seinfeld Strategy," we, too, can add emphasis and drama to even small, mundane points by raising our voice or feigning outrage, surprise, or exaggeration.

"Guess how many times she said goodbye before leaving? THREE TIMES. She said goodbye THREE TIMES before she actually left!" (Compare that to: "She's weird. She said goodbye three times.") Making a big deal out of a small detail is just ridiculous enough to be funny. In conversation, you might fixate on a minuscule detail, or repeat something that doesn't seem like it deserves repeating, or get frothed up over something that others will be surprised that you care about—all in service of fun.

In his series *Comedians in Cars Getting Coffee,* Seinfeld uses the Seinfeld Strategy in real life. In the real world, cruising around with other famous comedians and sipping espresso, he obsesses over tiny details about the cars, the coffee, and the minutiae of comedic delivery—things most people gloss over. He tells Sarah Silverman, "You're so funny," but insists on saying it again with a different type of expression to convince her that he really means it. In another episode, he and Larry David obsess over the pros and cons of eating hot versus cold food for lunch; the acceptable amount of time and physical distance for holding a door open for a person walking in behind you; and how quickly pancakes lose their allure once they've cooled. He's right—specifics are funnier than generalities. The name Sainsbury's is funnier than *supermarket,* Costco is funnier than *store.* Zooming in on visceral but mundane details, especially with exaggerated paralanguage, can be very fun.

COMPARE AND CONTRAST

My friend Naomi Bagdonas, the comedian who co-authored *Humor, Seriously,* teaches a strategy she calls "compare and contrast." John Mulaney used the compare-and-contrast strategy in

> an opening monologue for *SNL:* "Look, fourteen years ago, I smoked cocaine the night before my college graduation. Now I'm afraid to get a flu shot. People *change.*"
>
> The trick is creating stark juxtapositions—to show how two apparently dissimilar things are actually the same, or how two apparently similar things (e.g., Mulaney at age twenty-two and at age thirty-five) are actually quite different. You can compare kids and adults. They both need to eat to survive, but one works mightily to incorporate more fiber, while the other would prefer to subsist only on Sour Patch Kids and root beer.
>
> Or you can compare millionaires in New York City and millionaires in San Francisco. They both have exorbitant bank accounts, but one wears suits, the other ripped jeans and hoodies. You can compare your behavior at work versus your behavior outside work—napping is terrific during both. *New Yorker* cartoons like to compare modern-day humans and ancient cave dwellers . . . people and animals . . . locals and tourists. Compare and contrast is a great trick for bringing out the delightful absurdity of life.

As we sift through humor moves, remember that attempting humor takes tremendous courage. If you're going to do it, *do it*. Hesitance is palpable.

But it's important to find your style. To start, think about what you yourself find funny. What's your own comic perspective? What makes you laugh? Are you drawn to irony? To slapstick? To wit? To silly music or satire? To raunch? Everyone's taste is a bit different—which can be daunting as a speaker, but also freeing. You can find your niche as a humor creator, too. If you're uncomfortable making jokes in a large group, stick to cracking wise in one-on-one conversations. If you tend to be more serious and sensitive when talking one-on-one, try livening up your texts by sending *Onion* headlines to your friends, or sprinkling your emails with observational asides. A subtle, fleeting moment of mildly funny or lighthearted humor, a well-timed pause, an assertive laugh, a subtle slurp, a purposefully dramatic facial expression, even an honest email sign-off can work wonders to save us from conversational doom spiraling. *With love and licks, Alison.*

These moments of courage are rewarded even if they flop. Our research found that even when others perceived a person's jokes as unfunny or inappropriate, their views of the joker's confidence rose because they were *brave enough to try.*

If you still don't think you can land jokes or are too nervous to try, *get with it!* Just kidding—it's okay! You're amazing! Not everyone is meant to be hilarious, just as not every humor attempt will land as perfectly as you'd like. You can't control what people think. But you *can* control your follow-up. If you're worried that your levity attempt has killed the mood, label it a flop, right then and there. My funniest friends have go-to recovery jokes like "I'll be here all night" and "You're right, I'll cut that from the Netflix special" to acknowledge a flop and help everyone move on. It can be so endearing. Failing and recovering can be better than not trying at all.

And if you don't think you'll ever be funny, don't worry! It turns out that you don't even need to be funny to make people laugh.

The Best Medicine

The psychologist Robert Provine, a "bearded, congenial, and wisecracking presence" whose humor "could straighten the spine at its wicked best," according to his obituary in *The New York Times,* was the foremost expert on human laughter until his death in 2019. Provine liked to say that he investigated laughter as an alien might, asking: "What would the visitor make of the large bipedal animals emitting paroxysms of sound from a toothy vent in their faces?"

For a full year, Provine and an army of research assistants and graduate students spread out across Baltimore to listen in. They hovered in local shopping malls, patrolled city sidewalks, and hung out in student unions and other public spaces, searching for moments of lightness: peace, joy, bonding, tickling, farting . . . laughter. They captured over two thousand cases of naturally occurring laughter, paying special attention to the comments and behaviors that *preceded* it. And what they found has a lot to tell us about what actually makes us laugh and why we laugh at all.

Provine's team found that only 20 percent of laughter occurred in response to jokes, stories, or pranks. Instead, laughter followed banal remarks like "Look! It's Andre," "Are you sure?" and "It was nice meeting you, too." Provine wrote, "Even our greatest hits, the funniest of the pre-laugh comments, were not necessarily howlers: 'You don't have to drink, just buy us drinks' or 'Do you date within your species?'" It turns out that laughter actually serves many purposes other than expressing mirth, such as filling pauses, expressing awkwardness, showing politeness and deference, and smoothing interactions.

We see evidence of Provine's point in the speed dating transcripts. Many of Tom and Cassie's laughs, later in the conversation, were prompted not by jokes, but by their acknowledgments of awkwardness and game-playing. Tom's jokes near the beginning of their chat weren't particularly funny. But that's the key: they didn't have to be. The attempt showed that Tom cared enough about the conversation, and about Cassie, to *try*.

This was true for most of the speed daters. Another pair laughed even while they were discussing very serious topics—women's rights, "abortion and stuff," a "political philosopher from MIT," and how philosophers should "be more involved, make changes in the way people live." Each rated the other as a 2 (out of 10) on funniness. It sounds grim, until you learn that they also found the conversation quite enjoyable (both rated it a 9 out of 10 on enjoyment), and that both wanted a second date. It's clear their laughter served more complex purposes than simply to say "I think you're funny" or "This conversation is hilarious."

Authentic moments of belly laughter are magical. We cherish them and seek them. But we can't *really* make them happen very often, on purpose, during live conversation. It's more like the stars need to align—with the right people, on the right topic, and the right mix of weird circumstances. (They're probably more likely to happen with people we know well because we feel safe to let our guard down, play, and laugh with unfettered abandon.)

This is one of the big ideas at the heart of levity—it doesn't mean we have to be funny per se. Instead of seeking uproarious moments of laughter, we can simply prioritize moving toward *engaged enjoyment*, even if the steps are small. We can aim for everyone to be energized—excited to be part of this rousing conversational endeavor, together.

Positive energy can come in many forms, such as tone and pace of voice, affirmation, or nonverbal signals that trigger positive sentiment—and all can happen regardless of what topic is being discussed. A fleeting quip, a compliment, or an odd remark can work wonders. A creative digression can inspire, even if you must switch back to the heavy stuff eventually. Even a simple change in the surrounding environment can add a dash of levity. Researchers who study emotion regulation call these approaches "situation modification," achieved through adjustments to a context: you stand up, sit down, walk outside, change partners, stoke the fire—anything to shake things up.

We can focus our topic management skills squarely on the crucial pursuit of Levity. Rather than focusing on "being funny," instead think: *How can I steer topics to keep things energized and engaging? How can I make this fun?* There are many answers to these questions, and luckily, we don't have to be funny to make things fun.

CALL-BACKS

In my class, I ask my students to identify their favorite person in the world to talk to, initiate a new conversation with them, and record it. Analyzing the transcripts, we found that almost all these favorite talk partners use *call-backs*—explicit references to topics they've discussed together in the past.

Two of my friends once got a spam email for a million-dollar vacation package to Aruba and, listed among the perks included for that whopping price tag, was a "welcome beverage." When they talked about the email, they practically died laughing over this line—the idea that a million dollars got you one paltry free drink was just too much to handle. That conversation—and the uproarious laughter it provoked—stayed with them. Years later, it still comes up in their conversation. For example, one of them recently invited the other to a ticketed fireside chat between two authors. His friend paused, then asked, "Does it come with a welcome beverage?" The line is a gift that keeps on giving.

Call-backs are a form of levity because they lighten the mood and spark joy—and they often trigger authentic laughter. In one study, my fellow researchers and I asked people to build a time capsule of the mundane details of a day in their life—what songs they listened to, people they talked to, topics they discussed. Opening those time capsules weeks or months later was surprisingly lovely—the joy of rediscovery. We tend to remember extraordinary things that happen in our lives, while the ordinary, mundane details slip through the cracks of our memories. Call-backs are conversational time capsules. They help us rediscover the details of our shared past. Even if they weren't that interesting or exciting the first time around, revisiting them together can be extraordinary.

Master call-backers can reference things discussed earlier in the conversation, whether those topics were discussed a few moments ago or hours earlier—an impressive feat of listening, care, wit, and recall.

But they also call back to things covered in earlier conversations. Long-term call-backs are a hallmark of meaningful relationships, and we can achieve them with brief reflection before our conversations begin. *What did you talk about last time? What was your partner planning to do the last time you spoke?* You don't have to wait for them to raise those topics—*you* can call back to them, unprompted.

A STABLE OF STORIES

It turns out that go-to topics aren't just great for small talk. When my colleague Timothy McCarthy teaches public speaking, he encourages students to develop a "stable of stories," go-to anecdotes that they can practice in many different contexts.

We should develop our stories, he argues, the way racehorse owners select the horses to add to their stable. If you find that certain stories consistently make people laugh, use them again. If you're telling a story well, you'll be "co-narrated": listeners will interject, ask questions, laugh, and give back-channel feedback like *yeah* and *uh-huh* and *no way!* Consider emailing or texting your stories to a friend—writing them down can help you hone and memorize them. But if you try telling a story a few times, and people don't seem that into it—you don't get a lot of laughter, co-narration, or follow-up questions—then retire it. You want to keep only your strongest horses.

WEIRDNESS IS UNDERRATED

Great conversationalists, when they feel a conversation has become stale, often switch to off-the-wall topics to keep the energy alive. They can even do so as a pseudo-experiment to see what happens. My friend Mike switches to weird topics constantly. Once he was sitting at a dinner table where researchers were discussing the details of an issue no one really cared much about. At one point he said, "You know what I was thinking about on the walk over here?" Chuckling to himself, he said, "The homunculus. Remember that crazy diagram of the brain that shows how sensitive each body part is? Like the lips and fingertips are huge? I was wondering . . . if you had to remove a body part from your homunculus, what would you choose?"

I mean, so weird! But also fun and memorable, even several years later. Off-the-wall topics can come from anywhere—a strange thing we've read about, a song we've heard and loved (or hated), a tidbit from the news that doesn't make sense. It usually starts with "I saw this amazing thing . . ." or "Did you see that . . . ?" or "That reminds me of . . ." or "You know what I've been thinking about?"

We hesitate to get weird because we feel guilty for making digressions, as if they're disrespectful or will lead us too far afield. But—just like asking sensitive questions or raising sensitive topics—we can't know until we try. Minor digressions, especially to weird topics that you suspect your partner might like, are almost always good for enjoyment and can be extremely helpful for reaching new creative heights.

The challenge is timing. If you do need to get back down to business—to return to an important, if boring, topic—then spinning into the stratosphere and failing to accomplish your other goals is a mistake. But perhaps the bigger mistake is sticking so closely to mundane topics that everyone's too bored or disengaged to make progress at all.

How to Laugh

So far in this chapter we've focused much of our conversational striving on how to create levity. That makes sense, because levity can be too rare and creating it can be fraught. But perhaps the best and easiest thing we can do in practice comes from a different vantage point: to *support* levity, wherever it comes from. It's so easy: When someone else has the courage to make a joke—laugh. When you hear someone else laugh—pile on.

Hearing laughter from a partner evokes positive affect—it's a powerful signal of social connection and enjoyment. That's why Tickle Me Elmo is such a success. Elmo's laughter and joy make us smile and laugh back, even though Elmo has no feelings at all. (That we know of. I'm told scientists are still working on it.) Like Elmo, we should all laugh more. And if you can, go all in. One scientific paper dutifully reports, "Voiced, songlike laughs are significantly more likely to elicit positive responses than variants of laughter, such as unvoiced grunts, pants, and snortlike sounds." You heard it here first: don't just grunt or pant—laugh out loud.

The saying-is-believing effect for laughter is *powerful*. (And the *absence* of laughter is a conversation killer.) Just think of all those canned sitcom laugh tracks—they're there because laughter is wonderfully contagious. Provine, our laughter guru, even suggests that the people we think of as "funny" may just be those who spend more time laughing, which makes their partner more likely to laugh, too.

My own studies of laughter—observing people chatting while writing songs, getting to know each other, negotiating, and dating—corroborates Provine's findings: over 70 percent of laughter occurrences served purposes *other* than expressing mirth. Laughter smooths conversational flow by filling gaps and silences. It helps speakers manage the impressions they make—coming across as more affiliative and jovial. When I close my eyes and think about the person I want to be, she's laughing. And every once in a while, she's laughing so hard she snorts.

You Are the Best

A tall blond gentleman just inside the door was greeting guests as they arrived. His name was Dave.

"You're so beautiful, Alison. You're beautiful. You're beautiful," Dave said to me as I walked in. I smiled and thanked him.

Then he turned to my friend Kate and said the same thing. "You look gorgeous, Kate. Just beautiful. Beautiful."

Kate blushed and grinned. It was just the encouragement she needed to stride gracefully into the party. As she and I walked away, feeling jazzed from the compliment, we heard it again. "Beautiful. Just beautiful." We turned around to see a new set of guests arriving. Dave's script was on repeat, fading out as we got farther away: "You look gorgeous tonight. Just amazing."

Like a parrot who knows only a few good lines, Dave repeated his flattery with every person who walked through the door, and he didn't hide the audacious repetitiveness of his compliments. Repeated compliments were Dave's go-to move. I'd seen him do it, unabashedly, for years. But this was the day I realized the great power of flattery.

Even for people who suspected Dave was being insincere, his compliments worked. People loved him, and they respected him, too. I could see why: he made people happier. Later that night, I watched Kate kiss Dave. It seemed his compliments, however freely given, had worked their magic, even on my whip-smart friend.

Dave was on to something. His flattery warmed the room, lifting the mood through levity. Flattery is so powerful that we enjoy it even when the flatterer has obvious ulterior motives. Have you ever received a marketing mailing that says, "You deserve this" or "You're gorgeous"? Marketing scholars Elaine Chan and Jaideep Sengupta wanted to know if those compliments were effective. They randomly assigned people to read a plainly informative advertising message or a flattering version ("We are contacting you directly because we know that you are a fashionable and stylish person. Your dress sense is not only classy but also chic.") Their study participants understood that the firm's ulterior motive was to sell and that it had no actual personal knowledge of them, but this knowledge didn't subvert the good feelings that came from the flattery. The flattering messages were much more effective than the plainly informational ones in spurring sales.

We all want to be told we are smart and classy and chic, with exceptional taste. But many of us are shy about giving compliments, especially to people we don't know well. People think that giving compliments will make them seem incompetent, or put them in a one-down power position, or make their recipient feel embarrassed. To study this problem, psychologists Vanessa Bohns and Erica Boothby sent students out onto their campus to give random compliments. The students initially felt anxious that the compliments would make others feel uncomfortable and make themselves seem incompetent. But the study found that the students had *massively* underestimated the positive impact their compliments would have and overestimated how bothered and uncomfortable the compliments would make people feel. Even in hindsight, the students didn't understand how good their affirmations had made people feel—their anxiety was to blame. But as Bohns and Boothby showed, despite their anxiety, people who were nudged to give compliments felt much happier after giving them, and the recipients felt much happier after receiving them.

Like these students, flattery may be harder for you than it was for Dave, but you should still do it. Especially when you notice flattering things about others—things you'd love to hear others say to you.

Straight talk: I wouldn't advise you to say "You're beautiful" to everyone you see. (It's probably best to save comments on people's physical appearance for the people you're absolutely sure will be happy to hear them.) Instead, consider focusing on things you admire about them, like "I really admire how you handled that tricky situation," "You're brilliant," "I loved the graphic design in your slide deck," or "You're such an amazing role model for X because Y." You will likely get a smile and a closer friend.

Even if we aren't that funny, we have lots of tools to create levity in conversation. And, like Dave, we shouldn't worry much about seeming insincere. Everyone wants to feel delighted and admired. Also, as we'll see in Chapters 5 and 7, expressing flattery, admiration, and validation is even more powerful during difficult conversational moments.

No Regrets

One of Rachel Greenwald's assignments for her matchmaking clients (I use it for my TALK students, too) is to make a list of the traits they're looking for in a romantic partner, friend, or co-worker. It's a familiar ritual—one that many of us do instinctively during our single years or when we're looking to hire new employees. After Rachel's clients make their lists, she asks them to think of their favorite romantic partner, friend, or co-worker and to describe how those people make them *feel* when they're together.

The resulting lists, put side by side, offer a striking contrast.

Scholars and nonscholars alike often presume that people view others favorably if they match their ideal partner preferences. If Faye prefers kindness and towering height in a partner and Sonia prefers ambition and physical strength, then Faye should be especially attracted to tall, kind partners, and Sonia should be especially attracted to muscular, ambitious ones.

But this doesn't capture how partner selection, interpersonal attraction, friendship, and successful collaboration play out in real life. When you ask someone about their ideal romantic partner, friend, or colleague, their list of attributes may or may not include "funny," "silly," or "positive." But when you ask people how their favorite people make them feel, the list is always topped by one thing: "happy." We all prefer being with people who lift us up—people who make us feel happy.

If we checked in on our speed daters today, what would we find? Let's imagine that Tom and Cassie are still together, married and living in Austin, Texas. What do their conversations sound like now? Are they still flirting and teasing and gamifying? Have their levity habits changed to reflect the cozy closeness they've developed over the last seven years?

Let's fast-forward even further, as one decade passes and then another. When you look at what happens *later* in a relationship, research shows that levity is *crucial* for its long-term health and well-being. When psychologist Jennifer Aaker asked people on their deathbeds what they regretted most about their lives, the most common answer was "Wishing I had spent more time laughing with the people I love." Her work reminds us that a life (and career) *devoid* of levity is a far worse prospect than the moment-to-moment risk of the occasional flop. Years from now, when the speed daters we studied reminisce about their silly first date, recorded in the nerdiest of circumstances, I hope they exchange a knowing wink, clink the glasses of their welcome cocktail, and laugh and laugh and laugh.

THREE KEY TAKEAWAYS FROM CHAPTER 4

L Is for Levity

- **Find the fun,** rather than trying to be funny.
- **Give compliments** effusively.
- Don't just grunt—**LAUGH!**

CHAPTER 5

K Is for Kindness

IN JUNE 2019, THE AMERICAN painter, author, actress, and clothing designer Gloria Vanderbilt, an heiress of the great Vanderbilt shipping and railroad fortune, died of stomach cancer at age ninety-five—after a full life lived almost entirely in the public eye. Mourned as a "legendary fashion icon" and "early developer of blue jeans," she was buried next to her late husband, the author Wyatt Cooper, and her son Carter, who had died in 1988 at twenty-three.

Left to grieve was her youngest son, Anderson Cooper. The CNN anchor, then fifty-two, was no stranger to loss: his dad had died after open-heart surgery when he was only ten; his older brother had died by suicide when Anderson was twenty-one, leaping from the fourteenth floor of the family's New York City apartment building. These early experiences with tragedy, in many ways, equipped Cooper for his career. He was known for his cool level-headedness in difficult situations, whether covering war-torn regions like Somalia, Bosnia, and Rwanda or reporting on natural disasters like Hurricane Katrina and the 2010 Haiti earthquake. But something about losing his mother unmoored him. And so, two months after her death, he invited onto his CNN show another TV personality who he thought might offer some comfort: Stephen Colbert.

The youngest of eleven children, Colbert rose to media prominence by way of improvisational theater. He started as Steve Carell's understudy at Second City Chicago and later gained critical acclaim as a caricature conservative political pundit on Jon Stewart's *The Daily Show* on Comedy Central. Colbert's television antics might have made him seem an unlikely guest for Cooper (a harder-hitting reporter), but Colbert, too, was well acquainted with grief. He had lost his father at age ten in a plane crash, along with two brothers, Peter and Paul. More recently, he had also lost his mother.

The conversation between Cooper and Colbert—which has since amassed over two million views on YouTube—is one of my favorite public exchanges. Cooper explains that he wants to discuss grief and loss because he has found comfort in sharing his pain with others; he also wants to learn how tragedy has shaped Colbert. The two men, both wearing snappy suits, sit in director's chairs in Colbert's private office behind the scenes at *The Late Show*. It feels more like a conversation than an official interview, particularly because it's not conducted in front of a live audience. It's just two people talking.

Cooper opens by mentioning a condolence letter that Colbert recently wrote to him, then remarks on how tragedy creates a chasm between the person you once were and the person you are now. "To this day, I mark time between when my dad was alive and afterward, like the new year zero," he says. "It's like when Pol Pot took over Cambodia."

"Without a doubt," Colbert agrees immediately. "Without a doubt." He takes a deep breath. "There's another Steve . . . that kid *before* my father and my brothers died." In his memory, those years have taken on a ghostly quality.

"Like shards of glass," says Cooper. "Flashes."

Over twenty minutes, the two reflect on how loss has shaped their personalities, their careers, and their relationships. Cooper became a catastrophist, taking all manner of survival courses, while Colbert escaped into science fiction and fantasy. Cooper plunged into things that scared him; Colbert embraced awkwardness and humor.

The conversation is striking for its sincerity, its self-awareness ("I'm a WASP—we're taught to push our emotions deep down inside," says Cooper), and for the way these two grown men speak lovingly about their mothers.

But what makes the conversation as powerful as it is? It's how Cooper and Colbert show each other *care.* Cooper has clearly prepared for the conversation: he has read other interviews Colbert has given, can quote his words, and has notes and questions at the ready. Colbert is self-deprecating, and his flashes of humor are gentle; he allows Cooper to lead. When one man offers a vulnerable thought, the other nods, building on the idea or responding with a disclosure of his own. (Colbert jokes that he raised his mother instead of the other way around. "I completely understand that," says Cooper. "I always viewed my mom as a space alien. I had to protect her and show her how to live in this world.") They lean in, meet each other's gaze, and offer small encouragements ("Right, right," "Exactly").

And then about ten minutes in, there's a shift in their agreeableness.

"This is going to sound weird," says Cooper. "But I wish I had, like, a scar."

"Harry Potter!" says Colbert. "I know what you mean."

"Well, more like a Bond villain," says Cooper. "Running down my eye and my face." He's imagining a mark that signals to everyone he meets that "I'm not the person I was meant to be."

Colbert shifts in his seat. He does not agree with this idea. He takes off his glasses and smiles. "But you're *entirely* the person you were meant to be," he counters.

At this juncture—as the two men diverge on an existential question—their mutual care becomes especially noticeable. Cooper wonders if tragedy has made him a warped version of himself, and Colbert, a practicing Catholic, gently challenges this thought. "My experience of my particular faith—extremely imperfectly, admittedly—is that there *isn't* a different timeline," he says. "This is it. And the bravest thing you can do is to accept with gratitude the world as it is."

Cooper, who is famously private about his own spiritual beliefs, pushes back. He consults his notes. "You told an interviewer that you have learned to, in your words, 'love the thing that I most wish had not happened.'" He trails off, struggling with sudden emotion. "You went on to say"—here his voice breaks—"'What punishments of God are not gifts?'" His eyes redden, and he blinks back tears. *"Do you really believe that?"*

It is a charged moment. Colbert is quiet for a few seconds but does not hedge, dodge, or distract. After a pregnant pause marked by his unwavering gaze, he replies simply, "Yes." He is grateful, not for the tragedy itself, but for how he has suffered, so that he can connect with others and love them more deeply. So he can really understand what it means to be human. "I want to be the *most* human I can be," he says, "and that involves . . . ultimately being grateful for the things I wish didn't happen, because they gave me a gift."

It's probably not surprising that, in the weeks after the exchange aired, it went viral. Thousands of people responded online, sharing their own stories of grief. *The Washington Post* named it 2019's "best TV moment." Viewers commented on the interview's humanity and its humility, but many were especially moved by Colbert and Cooper's generosity toward each other. "It's rare to see the subject of an interview be there for the person who is interviewing them," noted one person on YouTube. "Colbert's single *yes* left me in tears," wrote another.

The conversation had a lasting effect on Cooper. Three years later, in 2022, he began a podcast about loss and grief, *All There Is*. Within two days of its release, the first episode reached the top of the podcast charts, downloaded more than four million times. The comfort that Cooper first found in his conversation with Colbert was something he hoped to offer others. "This is something we all go through," Cooper told *The New York Times*. "And this idea gave me strength, that I'm on a road that has been traveled by everybody, in one form or another." On the podcast, he reflects on the profound ramifications of his exchange with Colbert. "Stephen's words blew my mind," he said. "And I've been thinking about them ever since."

The Goals of Others

Throughout their conversation, Cooper and Colbert give a splendid exhibition of topic management, curious question-asking, and fizzy levity. While their talk mainly centers on the topics of loss and grief, along the way, they touch on music, religion, childhood, identity, siblings, Cambodia, poetry, survivalism, fear, elves, and mothers (as teacups and actresses and buddies and space aliens). Their topic landscape is vast and rich. They ask questions—fourteen questions in twenty minutes—like "Can you love your enemies?" and "Can you dig it? Are you hip to that scene, Daddy-O?" They laugh warmly every two or three minutes, following gentle jokes like "Instead of having an acting career, she started her own theater company, which was eleven children."

Still, it's their *kindness* that makes the whole thing feel more like a hug than a conversation. They meet each other's gaze, trade stories, listen, laugh, and affirm. Behind their behavior is an invisible, but palpable, focus on the other person—their words, their struggle, their life, their needs. It's this focus that not only elevates the conversation but, to my mind, allows them to achieve their conversational goals. Cooper is driving the conversation—his goal is more evident on the surface, seeking some kind of insight that will help him process his active grief. And he seems fairly sure that he will find it, which may be why he is filming the conversation to share with a broad audience. At the same time, he is aware that he's asking Colbert about one of the worst things that's ever happened to him.

Cooper is careful to create an environment in which Colbert feels comfortable and supported while describing his own tragic past, and he seems to sense that Colbert's goal is to be vulnerable as a way of offering empathic instruction. Even Colbert's pushback is a way of supporting Cooper's need for support, by gently insisting that Cooper (and anyone else watching) might consider a different way of thinking about the tragedy of loss.

Kindness, the last, most challenging, and in some ways the most essential of the TALK maxims, boils down to a simple premise: trying our utmost to put others' conversational needs first. Yes, you're right, that sounds hard, and it may be impossible. *Always* putting others' needs first is unrealistic—and not always optimal.

But even if kindness is an ideal that we can't achieve all the time, reaching for it relentlessly is the best chance we have of being the people we mean to be. Trying to prioritize others requires a continuous (and sincere) devotion to our partners. The kind conversationalist's job is to figure out what their partner needs. And whatever they need—encouragement, hard feedback, new ideas, a quick laugh, a sounding board, challenging questions, a break—kindness means helping them get it.

While topics and questions provide the substance of the coordination game, and levity keeps the players interested, kindness creates a space where the game can flourish—where the players feel respected and valued. In order to win, each player needs to focus on doing whatever they can to help the others play their best. Topics, Asking, and Levity help us decide what to say, but Kindness takes these skills to an elite level. Kindness is the master class.

Kindness Takes Work

Practicing kindness seems like it should be simple. It's like saying, "Just be nice!" or "Do the right thing!" But in real life, we screw it up all the time. And the reason has much to do with our egocentric human instincts.

The Swiss psychologist Jean Piaget, grandfather of developmental psychology, was the first to trace the development of cognition in children. Around age seven, as they move out of a state of "extreme egocentrism," children begin to recognize that other people have different perspectives and needs. Until then, their behavior is marked by hilarious (and at times, frustrating) self-centered flops, like not understanding that other people can still see them when they put their hands over their eyes during a game of hide-and-seek, or that shouting "That's mine!" or "I was using that!" isn't compelling justification to hoard a toy or snack for themselves (and is more likely to incite conflict than quell it). But just how well do we shed our "extreme egocentrism" after age seven? The answer is: not very well.

It's a devastating fact that affects every aspect of conversation. Egocentrism is always operating in the background, a dark puppeteer that sabotages our ability to converse effectively. It sabotages our ability to choose good topics because it makes us choose ones we like (on the assumption or hope that our partner will like those topics because we like them). It sabotages our ability to ask good questions (especially ones our partner would be excited to answer). It sabotages our ability to create moments of levity and humor that would make our partner feel pleased and engaged. The adult mind is built remarkably well for survival but less well for thinking about others, especially on the fly, and especially when their preferences and goals differ from our own.

Recent studies by the psychologist Boaz Keysar show how egocentric we are during conversation. Keysar and his coauthors asked pairs of people to sit back to back, taking the roles of speaker and listener. The speakers were asked to imagine a scenario like *You suspect your friend has been planning a surprise for you.* Then they were told to deliver a conversational line related to this scenario (with any inflection of their choosing). In this case, the required line was "What have you been up to?" This is the type of question that might cause the speaker to squint their eyes, raise their voice and feign cheeky skepticism, like an old-timey inspector or investigator. Such nonverbal and acoustic cues may help listeners divine their intended meaning (that they suspected a surprise was in the offing) from an infinite set of possibilities.

In Keysar's experiment, listeners were asked to identify the speaker's intended meaning from a predetermined list of four plausible options, including the correct option, which was known only to the speaker. The listener who heard the question "What have you been up to?" could choose from: "The speaker is 1) suggesting that I've been unfaithful, 2) suspicious that I've been planning a surprise for them, 3) angry that I'm thirty minutes late, or 4) curious how I've been doing recently."

So what was the result? People vastly overestimated their shared understanding. Listeners thought they had identified the speaker's intended meaning (option 2) correctly 85 percent of the time, while speakers guessed that the listeners had understood their intended meaning 70 percent of the time. The listeners actually understood the speaker's intended meaning only 44 percent of the time. That's really low, and really off.

Keysar and his coauthors didn't stop there. Perhaps their most surprising finding came when they repeated the experiment with people who *didn't speak the same language*. Speakers made statements in Chinese to listeners who only spoke English, yet both speakers and listeners *still* overestimated their understanding. American listeners believed they had understood the speaker's intended meaning—conveyed by their tone of voice—65 percent of the time, and Chinese speakers believed the American listeners had understood them 50 percent of the time. Though their confidence was less than that of the participants who spoke the same language, they were still way off: the American listeners correctly identified the intended meaning only 35 percent of the time.

The Irish playwright George Bernard Shaw is said to have described this conundrum: "The single biggest problem with communication is the illusion that it has taken place." Keysar said his studies showed the "extreme illusion of understanding" that speakers routinely experienced in real conversations.

When it comes to conversing with kindness, our problem is that we can't prioritize and pursue others' conversational needs if we don't understand them, especially when we mistakenly believe that we do.

Of course, it's impossible for us to be completely selfless—and we shouldn't be. We all have our own needs on the low-relational side of the conversational compass. But the more you can figure out the minds of others, the better positioned you'll be to respond to them.

In Keyser's experiment, the four possible meanings are very different: someone believes you've been unfaithful, or they think you might be planning a surprise for them, or they feel angry that you're late, or they're simply curious about how you've been doing. How you respond will influence how the conversation goes in the short term, and how your relationship will unfold over the long run (if it continues at all). In normal conversations, we have to divine others' meanings all the time. Suppose you're doing the dishes, when someone comes over and says, "I'd be happy to do the dishes." You have to interpret whether they'd actually love to do the dishes, or it's the last thing on earth they want to do, or something in between. Even if you accept their help, you're better positioned to express an appropriate amount of gratitude, crack a relevant joke, or give them a much-needed hug if you know how much they actually wanted to do the dishes in the first place. In a practical sense, even when we prioritize our own needs over our partner's, it's better to do so knowingly.

Colbert's behavior during the CNN interview was kind in many ways. Though the segment was framed around an interview of Colbert on Cooper's show (the title card is "The Stephen Colbert Interview"), Colbert quickly pivoted to shift the focus toward supporting Cooper. Colbert was uniquely positioned to understand Cooper's perspective and his goals—to realize that Cooper was hoping that Colbert's struggles with loss and grief (by this point many years in the past) might offer him support and solace in his fresh time of need. In a way, Colbert had been preparing for this conversation for forty-five years, ever since the loss of his father and brothers ("Dad and the boys") at age ten. In his own words, his experiences allowed him "to get awareness of other people's loss . . . to connect . . . to love more deeply . . . to understand what it's like." He figured out exactly what Cooper needed—support, reassurance, a new perspective on self-identity and loss—and he delivered it.

Respectful Language

To be kind like Colbert and Cooper, we must dig deep into the practice of conversational kindness, starting with its two fundamental components: how we speak (respectfully), and how we listen (responsively).

The first, most basic prerequisite for kindness is *respect*. We must respect and value other people in order to prioritize their needs. As children, we were all taught about the importance of respect in school and by the Queen of Soul, Aretha Franklin, in her legendary tune "Respect" (which, with all due respect, was written by Otis Redding). But even listening to this anthem on repeat doesn't give us a clear definition of what R-E-S-P-E-C-T looks or sounds like in everyday conversation.

It's not Aretha's fault, or your primary school teacher's. Respect is a complex, gradient concept that can be enacted through an infinite number of thoughts and behaviors. We know it when we see it— Cooper and Colbert's mutual respect is tangible—and we know what egregious behaviors rule it out. But what exactly do highly respectful people say and do from one conversation to the next? Is there a predictable language of respect?

The science of conversation helps us go beyond our vague, commonsense understanding of respect and drill down to the *language* we actually use. In practice, our language has tiny, specific elements that communicate respect (micro-kindnesses) that we should use more often in our attempts to be kind. And certain elements of our language communicate disrespect (micro-aggressions) that we should use less. An amazing study by a team of computer scientists from Stanford University aimed to identify exactly those elements.

The Stanford researchers examined body camera footage from 981 traffic stops conducted by 245 different officers from the Oakland

Police Department during the month of April 2014. According to Oakland Police Department policy, officers turned on their cameras before making contact with a driver and recorded the encounter for the duration of the stop. From 183 hours of recorded footage, the researchers obtained more than 36,000 officer utterances.

Their dataset is remarkable for its sensitivity—these interactions are high stakes for police and civilians alike—and even more for its scale. It was the first time anyone had been able to measure, rigorously, how police officers communicate with the public. These interactions—in a very specific and potentially fraught context—yield insights that we can apply to our everyday conversations, too. The kind of language that, coming from officers, makes citizens uncomfortable during a traffic stop is likely to make *your* friends, your romantic partner, your mom, and everyone else uncomfortable in less charged circumstances, too.

The researchers used different computational tools to sort all 36,000 officer utterances into twenty-two categories, each given a different linguistic label (see the graph on page 129). For example, whenever officers used *sorry, oops, whoops,* or *excuse me,* as in "I'm so sorry for the mixup, that's my fault," the researchers labeled it as Apologizing. And anytime an officer used *not* or *never,* as in "That's not good," they labeled it as Linguistic Negation. The researchers ranked these features in terms of how respectful they were—which ones made the officers come across as more and less respectful to real citizens in California? They plotted features according to the degree of respectfulness they convey: the most respectful ones are at the top, while the least respectful are at the bottom.

What's astounding about these findings is that these computer scientists broke down what respect means, not only to Aretha but to 981 real citizens in California. For our purposes, their groundbreaking analysis reveals three commonsense rules about respectful language that we can apply in our everyday lives: respectful language makes people feel (1) seen and known, (2) good to be with, and (3) worthy of care.

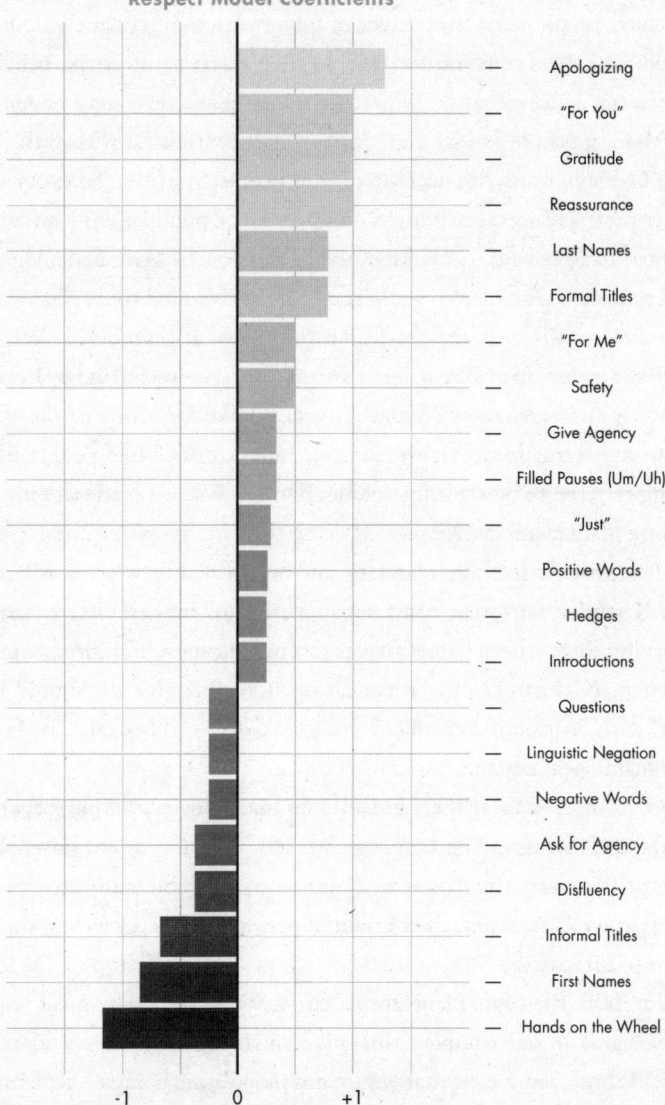

SEEN AND KNOWN

To start, people want to feel seen and known—to be correctly understood. The kind conversationalist, in their quest to prioritize others' needs, can assume that every partner wants to feel seen and known.

Making people feel seen and known starts with what we call them. The Oakland study found that calling people by their names (or by other preferred forms of address) matters tremendously.* The citizens wanted to be seen as individuals, not as sheep to be herded, managed, and punished. In order to feel respected, they wanted to be addressed personally and with the formality the occasion demanded. While terms of endearment like *honey* or *bro* may be quite warm and welcome in some contexts, more formal references, like *Sir, Mrs.,* or *Dr.,* are more respectful in others (including traffic stops). Most people like being referred to by their first name; Barbara Walters made a point of calling Ms. Lewinsky "Monica" because their interview required Walters to provide feelings of safety and warmth. Likewise, I ask my TALK students to wear name tags for most of the semester because knowing each other's names makes interacting more comfortable and familiar. Nicknames can be endearing, too, though they should be used with caution if someone doesn't yet feel close enough to be on a first-name or nickname basis.

Of course, making people feel seen and known goes far beyond using their names. The language we use—the topics we raise, the questions we ask, the stories we share—can touch on many aspects of our partners' identities, their essential personhood, and we must do so with sensitivity.

In 2006, the Nobel laureate Amartya Sen wrote, "The main hope of harmony in our troubled world lies in the plurality of our identities." I think Sen meant that we are not monolithic beings—we aren't

---

* In the traffic stop data, citizens preferred being addressed by their formal title and last name, as in "Good evening, Mr. Driver" or "Here's your license back, Miss Wheeler." And on the disrespectful end, the officers who used citizens' first name and an informal title—*dude, bro, boss, bud, buddy, champ, man, guy, brotha, sista, son, sonny,* or *chief*—were rated as more disrespectful.

only our race, only our gender, or only our religion. Rather, each human is composed of beautiful, diverse aspects: race, sex, gender, nationality, language, age, religion, education, wealth, values, ideologies, sexuality, health, family, friends, hobbies, experiences, preferences, and so on. Every conversation partner is different from you, even your close friend who seems so similar to you. My identical twin, who is an exact copy of my DNA and upbringing, possesses talents, habits, preferences, and passions that I do not. Of course, some partners are more different from you than others (or at least *feel* more different from you than others). Your goal as a kind conversationalist is to walk beside your partner, no matter who they are, and work to make them feel seen and known.

GOOD TO BE WITH

The second type of respectful language revealed to us by the body camera study is emotional tone. The study found that citizens perceived words with a positive sentiment like *good, great,* and *wonderful* (the linguistic feature Positive Words) as respectful, while they perceived words like *bad* and *worse* (Linguistic Negation and Negative Words) as disrespectful.

Positive and negative language is important because the language we use evokes commensurate feelings—good and bad—in our partner. Positive words give a sense that things are good, while negative words evoke unpleasant feelings. This simple point is important because people yearn to feel that they are good to be with. Everyone wants to feel that their conversation partner is happy and comfortable in their presence, not annoyed, uncomfortable, or eager to leave.

These findings serve as a reminder that emotional tone—and levity—matter tremendously. Our exchanges need to be enjoyable, not just for fun's sake but also to convey dignity and human value. Making our partner feel good is respectful because it shows that we enjoy being together—that we want to give them our attention—while making them feel bad is disrespectful because it makes them wonder if we'd prefer to be alone or with someone else. "It's so nice to hear the sound of your voice," "How kind of you to ask," and "I'd love to hear about your weekend" convey respect.

WORTHY OF CARE

In the traffic stop study, officers' use of certain linguistic features—apologizing, saying "for you," expressing gratitude, issuing reassurance, mentioning safety, giving agency*—sent an important message: that the officer believed the driver was worthy of care.† All of us want to feel that we are worthy of others' attention and care. That we are valuable. That we matter. Tiny aspects of our language—things like "Oh, I'm so sorry" or "I appreciate you" or "You're doing great" or "No worries, you're good" or "Be safe!" or "Do you want a snack for the road?"—convey that others' health, safety, comfort, and happiness matter.

During traffic stops, the three categories of respectful language—making people feel that they are *seen and known*, *good to be with*, and *worthy of care*—even when conveyed through very subtle linguistic choices, were consequential. In the study, not only did disrespectful language correlate with more vehicle searches and citations, but expressions of respect or disrespect in personal interactions with police officers played a central role in citizens' judgments of the procedural justice of the police as an institution, and they affected the community's willingness to support or cooperate with the police overall. Perhaps it is no surprise that the central finding of this research is that police used more respectful language with white drivers than with

---

* Citizens rated Apologizing as the most respectful linguistic feature of all. The use of Apologizing language signals that the officer realizes he makes mistakes, too, or acknowledges that it's an unpleasant situation for the citizen. We'll revisit the remarkable power of apologies in Chapter 8.

† Officers' use of the least respectful linguistic features—including Hands on the Wheel and Ask for Agency—conveyed that the officer didn't believe the citizen was worthy of care. As Hands on the Wheel is a go-to line in dramatic movie scripts, it's perhaps no surprise that it was dead last in the respectfulness ranking.

Black drivers. This finding is acutely devastating. It provides evidence, in small linguistic choices, of the pervasive disrespect that can make whole populations feel like they need a social movement to convince others that they matter.

The Contagion of Disrespect

While respectful language can humanize us and pull us closer, subtle markers of disrespect can dehumanize us and tear us apart. It's very difficult to be on the receiving end of disrespectful language (or the noticeable omission of respectful language) and hear anything but alarm bells. Receiving even subtle expressions of anger or disrespect floods the amygdala—the small, almond-shaped fight-or-flight response center in the brain—with neurochemicals as if we're in physical danger. This process, referred to as amygdala hijacking, overtakes rational brain functioning. It makes us feel anxious, which triggers a freeze-or-flee response or makes us feel reciprocally angry, triggering a fight response. Freezing, fleeing, and fighting all foreclose the possibility of conversational kindness. The problem is that we may not perceive our own language as disrespectful because our linguistic choices are subtle and often operate beneath our awareness.

Psychologists Trevor Foulk and his coauthors asked working adults about their experiences with "low-intensity negative behaviors" at work. Their findings show that rudeness and other negative microbehaviors spread "like the common cold": they're widely dispersed, easy to catch, and anyone can spread them. These scholars focused on how people form logical connections between others' behavior and harsh punishment. In one example, a worker arrives late to a meeting and is chastised, first with very harsh language: "How can you be this late? What is wrong with you? We started over fifteen minutes ago! Look at you. . . . I do not know how you expect to hold any sort of job in the real world with this type of behavior, but we've already started. It's too late now, I need you to leave." In a second case, a late-arriving worker is addressed with more respectful language: "I'm sorry, but we've already gotten started, and unfortunately it's a little too late for you to start at this point. Email me later, and we'll see if we can find another session to get you into."

The onlookers may not have responded harshly themselves, but seeing the disrespectful version lodges it in their minds and changes how they think. It's not that it stays active within them forever, but instead, the next time they see someone show up late in a similar context, it will reactivate this language, and they'll be more likely to repeat the disrespect toward the latecomer. Luckily, these effects seem to wear off fairly quickly over time. If an individual experiences a rude encounter, followed by a nonrude encounter, the new episode overwrites the association between the conversation setting and rudeness. But, when considered across a whole social network, even onetime repetitions of disrespect are extremely damaging. It means we're all hearing them and repeating them. We're all catching the virus.

While our instincts to accommodate, mirror, and merge with others' spoken language are often good, in the case of rude language, these instincts are dangerous. We become less respectful not only when we feel disrespected but also when we simply hear others being disrespectful to someone besides ourselves. Simply *witnessing* an act of incivility or rudeness profoundly influences our mindset. Research by psychologist Kathleen Vohs shows that witnessing even one small act of disrespect or incivility in the morning colors our perception of the whole day. Seeing micro-expressions of incivility creates a "vicious cycle" of disrespect and cynical beliefs about human nature, predisposing us to be more disrespectful toward others. Our interactions—whether civil or uncivil—affect not only ourselves and our chat partner but everyone around us as well.

Disrespect spreads like salacious newspaper headlines printed and reprinted by our own mouths. But it also gets in the way when it comes to practicing kindness—even brief interactions that we witness between others can turn good people, unconsciously, into less-good people. Kind conversationalists must be on the lookout to stop the spread of disrespect by refusing to repeat it.

Of course, this move is particularly challenging when you are the victim of someone's disrespect. I once had an executive ask me, right before class, if it was worth his time to stay for my session. It was the first time I'd taught in the executive program at HBS, and I was the only female professor in the program. It was a vulnerable moment, and I was really mad—wounded, outraged. Did he think, based only on my appearance, that my session wasn't going to be useful? In a context where, as a young woman, I have to fight for respect, it seemed egregiously disrespectful. Still, I mustered my patience and said, "Yes, you should stick around—it's going to be great."

Some months later, I was lamenting about this encounter with a colleague who had taught in the same program. He remembered the executive and told me something I didn't know: that executive had been picking and choosing sessions to attend because he had recently been diagnosed with a serious illness and needed to rest. It wasn't about me at all. It was a reminder of just how little we know about others and their needs—as Boaz Keysar's studies of English and Chinese speakers showed, we never know how far off our understanding of basic conversation can be. In this uncertain world, where no one is fully known and our illusion of understanding can be quite extreme, it's best to give people the benefit of the doubt—to err on the side of respect—even when it's desperately hard.

Responsive Listening

Using respectful language is kind, and using disrespectful language is unkind. But our linguistic choices—the words we use—are only the tip of the iceberg when it comes to practicing conversational kindness. The Kindness maxim isn't just about talking—even more importantly, it's about *listening*.

For decades, experts and the popular press have espoused the importance of "active listening." This technique involves using nonverbal cues of listening—you smile, nod, laugh, lean toward your partner, meet their gaze, mirror their gestures. Those tips are extremely useful. People who are perceived as "good listeners" accrue all kinds of benefits. Romantic partners who are seen as good listeners have better marriages. Teachers who are seen as good listeners get better evaluations. Doctors who are seen as good listeners are subject to fewer lawsuits.

But *perceptions* of good listening are useful only to a point. Here, in the spirit of kindness, I want to emphasize the importance of actual listening—not just employing nonverbal cues to convey the impression of listening but doing the hard cognitive work of *really* listening, training our attention and brainpower on others. It turns out to be trickier that we think. In order to prioritize others' needs, we must figure out what they are. And the only way to figure out what they are is to pay attention and to process what we see and hear. Listening is the all-important glue that holds the TALK maxims together.

If active listening is expressed through *nonverbal* cues, my research—between thousands of strangers, people in close relationships, and doctor-patient pairs—shows that the best way to express actual listening is through *verbal* cues—asking follow-up questions, paraphrasing, giving credit to previous speakers, using call-backs to ideas from earlier in the conversation (or earlier in the relationship). While nonverbal signals like smiling and nodding give the impression that you're listening, which is important, they are also easy to fake—which can inadvertently give you an easy way out of actually paying attention. (Think of all those times you've found yourself smiling and nodding on Zoom, while your mind wanders elsewhere.) But *verbal* cues of listening—words that *show* you've heard your partner—are possible only if we've actually listened attentively. There can be no illusion of understanding if we use our words to express our actual understanding.

Verbal cues of listening are examples of *responsiveness*—the extent to which a listener expresses understanding, validation, and care. Responsiveness is what individuals mean when they report "feeling heard" by a conversation partner, and it's at the very heart of the kindness we're striving for. So while good conversationalists practice active listening, great conversationalists practice *responsive listening*. They constantly try to perceive, interpret, question, repair, and build on each other's ideas. And higher responsiveness to a partner strongly predicts the success of long-term relationships by improving intimacy, attachment, and emotional health.

Are You Even Listening?

Responsive listening takes work. Amid the onslaught of visual and auditory information that enters our eyes and ears from the world, processing the meaning of what someone is communicating requires a lot of focus. And focus is made doubly hard by our biology. Our minds were built to *wander,* so much so that some philosophers have suggested that a large portion of human conversation includes very little actual listening at all. The philosopher Abraham Kaplan, for example, coined the term "duologue" for a conversation in which neither party actually listens to the other. A duologue is more than a monologue because the participants take turns speaking, but it is not quite a dialogue because they don't really respond to each other—their conversation lines run side by side, but never quite intersect.

Our conversations may resemble duologues more than we'd like to believe. In one experiment using instant messaging, Bruno Galantucci and his colleague Gareth Roberts found that between 27 and 42 percent of participants did not notice when their conversation partner was replaced with a new partner, even when the new partner made obvious blunders, like incorrect references to their partner's gender. In another study, Galantucci found that when a researcher posing as a participant said "Colorless green ideas sleep furiously"* in the middle of a face-to-face conversation, only one-third of the participants noticed the nonsensical sentence, and 90 percent couldn't recognize it from a list afterward.

In my research with behavioral scientists Hanne Collins, Ariella Kristal, and Julia Minson, we've found that during conversations with strangers, people reported listening attentively 76 percent of the time and allowing their minds to wander 24 percent of the time. (They privately reported the attentiveness of their listening every five minutes throughout the conversation.) That's almost a quarter of the conversation lost to inattentiveness. And because admitting that you're not listening isn't socially desirable, that percentage probably underestimates how much our minds actually wander.

This might seem grim, but we also found a silver lining: people were able to nimbly adjust their listening when they were motivated to do so. They listened more attentively when we asked them to, especially when they would be paid based on how well they remembered what was said or on how well their partner rated their listening. (Money talked, they listened.) Responsive listening is hard work—but it is possible, especially when you care to do it. And to be a great conversationalist, it's essential that you care.

Back-Channel Feedback

If you've devoted the effort to listen as best you can, it's important to show it through responsive listening—not just by nodding, smiling, and laughing (cues that can be faked) but also by using explicit verbal signals of listening that can't be faked. In the CNN interview, Stephen Colbert persistently uses responsive listening to show that he's actually listening to Anderson Cooper. In the following exchange, where Cooper is talking about how people are hesitant to

---

* This grammatically correct but nonsensical phrase was coined by the linguist Noam Chomsky.

discuss death, Colbert's interjections are indicated in angle brackets and *italics*:

> It's always interesting to me, how, when you, you know, bring it up, meeting somebody for the first time, and they say, "Oh, I'm sorry to bring it up." *<Mmhmm.>* As if, what they don't *<Like you'll forget the person who died>* realize is I'm thinking about it all the time. *<Constantly, exactly.>* It is as you said, *<Exactly>* it is, it's one of my arms—it's an extension of who I am.

Colbert then quickly finishes Cooper's sentence: "And quite possibly, for the rest of your life."

Colbert's constant, subtle interjections are called *back-channel feedback* (or *back-channels* for short)—precisely timed expressions of verbal affirmation, encouragement, understanding, or confusion. His back-channels, along with his sentence-finishing habits, are examples of responsive listening. Back-channel feedback is part of the collaborative process of *grounding,* in which the speaker and listener coordinate their contributions to ensure their mutual understanding.

Psychologist Janet Bavelas and her colleagues did a micro-analysis of psychotherapy sessions and of conversations between strangers. They found that grounding is a three-step process:

- The speaker presents information. ("I do not love this cheese.")
- The listener displays some level of understanding. ("Oh yeah?")
- The speaker acknowledges (or corrects) the listener's display. ("I love most cheeses, but this mustard gouda is gross.")

Bavelas's work shows that listeners aren't just passive or quiet bystanders absorbing information. As we've seen in Chapters 3 and 4, conversational listeners are "co-narrators" of their partners' stories and disclosures through back-channel displays. Without back-channel feedback (*huh?*), it's impossible for a speaker to know if and when their partner is hearing and understanding their speech (like an online webinar where the speaker is talking into a soulless, silent, disorienting void). Back-channel feedback helps speakers sense when they may need to provide more information.

In addition to guiding information exchange through grounding, back-channel feedback can also add energy and fizz to a conversation (as we saw in Chapter 4) by actively encouraging and reacting emotionally to what the speaker is saying. Our subtle use of *oh* and *really?* and *no way!* cheer them on or tell them to go in a different direction.

I've Been Faking It the Whole Time

Like nonverbal cues, back-channels can sometimes be faked. The timing and rhythm of conversation is so instinctual that most people can sense when the conversation demands a generic *oh* or *mm-hmm* to keep it alive. In Bavelas's work, generic, disengaged back-channels tended to be more sparse than engaged back-channels. Worse, faking it can actually have the reverse effect from the one intended: speakers who received generic back-channels struggled to deliver their stories with panache, an interpersonal goal-stifling that surely qualifies as a kindness fail. On the other hand, specific, engaged back-channels tracked the logic of the content—like when Colbert said "Like you'll forget the person who died" or "Constantly, exactly," to confirm and emphasize the logic of Cooper's story.

As speakers, we need our listeners to react. A silent void is deadly not just on video calls but in person, too. And the positive effects of engaged listeners who act as co-narrators are recursive: good back-channels help punctuate what the narrator is saying, and they improve the narrator's delivery, too. Think about the last time you shared something important or personal or vulnerable with someone in a conversation. Were you monitoring their response as you spoke? Were you hungry for any shred of affirmation, acceptance, reassurance, or encouragement? Did you feel comfortable and safe as you spoke? What was your partner doing if you did? What were they doing—or not doing—if you felt uncomfortable?

As listeners, we don't want our partners to feel uncomfortable. Remember, we want them to feel that it is good to be with them. If we find ourselves going through the motions of giving back-channel feedback, humming sporadic *mm-hmm*s while our minds wander elsewhere, we should learn to recognize that this behavior isn't serving the conversation—it's a signal that we've disengaged and that our micro-lies of attentiveness are harming our partner's ability to engage, too. While back-channels are part of a collaborative process, noticing when we're doing it authentically versus mindlessly is a skill to be honed.

The Responsive Listening Toolkit

To be responsive listeners, we can use sincere back-channel utterances like Colbert's, and we have other options, too. We learned about call-backs as a tool for topic management and a reliable source of levity—joyful moments of nostalgia and rediscovery. (Yes, I'm calling back to call-backs!) But call-backs are also great indicators of responsive listening. They show that you've not only listened to something your partner has said but also *remembered* it and found it interesting enough to return to it. That is a supremely kind move.

The next tool in our toolkit is our old friend the follow-up question, that legendary superhero of question types. Follow-ups are another great responsive listening tactic. After all, we can probe for more information on a specific topic or utterance only when we've been listening to our partner in the first place. Follow-up questions steer the conversation while also expressing good listening—that's why they're so heroic.

Another responsive listening tactic is interrupting. What?! It's true—interruptions have a bad reputation. We tend to think of them as rude and disruptive—and of chronic interrupters as people we'd like to avoid. However, my research with Matteo Di Stasi suggests an important distinction. While off-topic interruptions—cutting a speaker off to switch to an unrelated idea—are jarring and abrasive, on-topic interruptions are actually a signal of engaged listening—a form of successful co-narration. On-topic interruptions suggest that the listener is so engaged and excited about where the conversation is headed that they couldn't wait until the end of the speaker's comment.

Finishing each other's sentences, as Cooper and Colbert do, counts as an on-topic interruption because there are no pauses between turns, and because one speaker picks up where the other left off. On-topic interruptions are a signal of good, bubbling discourse, engaged listening, and often a sense of closeness between the speakers. Some speakers may still find them a bit annoying if they're unable to finish their thought or if the listener has incorrectly anticipated what they were going to say. But on average, on-topic interruptions are a different (better) species than their hurtful and rude cousin, the off-topic interruption.

Back-channeling, asking follow-up questions, paraphrasing, lavishing credit, and on-topic interrupting are fantastic tactics we can all work on, but ultimately, verbal expressions of responsive listening can be *even simpler:* Just repeat or reformulate what has just been said. "Your mom said she doesn't eat meat anymore?" "Would it be right to say that you don't like politics, then?" "Did you say 'chillax'?!" Repeating or reformulating what you've heard helps you, for a moment, merge minds with your partner and makes them feel heard.

Active Responders

Once you realize that great listening is expressed through verbal response, it can make you listen *differently*. A person trying to be a responsive listener is always on the lookout for good fodder coming from the speaker—little moments (like a loathing of mustard gouda) that you can actively validate and affirm, points of vagueness that you can ask them to clarify, fragments that don't seem quite right, tidbits that beg for a silly follow-up joke, words and content a partner shares that surprise you and are worth noting, or a stumble over words that you can reformulate. All these examples are things your partner says or does that you could incorporate into what you say and do next. In this way, aiming to use these verbal expressions of listening may help you find conversation more engaging—like a more interactive treasure hunt. Responsive listening may increase the extent to which your partner feels heard and is actually *being* heard.

The best thing about responsive listening is that its benefits are not bounded by the limits of one conversation. Listening unfolds gradually over time as individuals take turns speaking and not speaking, both across the turns within a single conversation, and across multiple conversations within a relationship. We can express our listening at various points in time—in the moment (by providing backchannel feedback while someone is speaking), in the next turn (by directly acknowledging or paraphrasing what someone has said), several turns later (by calling back to something mentioned earlier), or in a separate conversation (by asking a follow-up question about something discussed yesterday or a year ago, or simply following up with an email to say "I loved when you gave Jean that compliment").

Verbal expressions of listening present unique opportunities. Consider a colleague who, in a separate email chain days later, gives you credit for a point you made in a team meeting a month ago. A friend asks you whether you enjoyed that concert you were looking forward to when you last spoke. A relatively new acquaintance asks how your leg is feeling because it was in a cast when you first met. Just as listening is fundamental to holding an effective conversation, responsive listening is a key ingredient in the secret sauce for successful relationships.

In the opening monologue of his hit podcast *All There Is,* Anderson Cooper refers back to the amazing conversation he had with Stephen Colbert years earlier. He's showing long-term responsive listening himself, just as Colbert showed responsive listening to him. Their conversation was so jointly constructed, it's easy to forget who's the interviewer and who's the interviewee. So disorienting is their dynamic that Colbert even pokes fun at it. At the end of the conversation, as if he's been the interviewer the whole time, he looks straight into the lens of the camera and, switching into his late-night TV voice says, "Anderson Cooper, everybody. We'll be right back."

Kindness in Practice

When I was in college, I met a talented drummer (and married him several years later). I've loved playing in rock bands ever since. As a singer and songwriter, perhaps it's no surprise that I've always wanted to study people's conversations while they write songs—it's a creative challenge that captures many difficult and beautiful things about collaboration. Taylor Swift once described songwriting as "this little glittery cloud that floats in front of your face, and you grab it at the right time."

In 2017, my glittery dream took a scholarly turn. Together with psychologists Jennifer Aaker, Maurice Schweitzer, and Brad Bitterly, I recruited 204 strangers and put them into pairs. In our behavioral lab, we assigned the pairs to write and record an original song in fifteen minutes—the amount of time Lady Gaga says it took her to write her smash hit "Born This Way." Each pair got their own songwriting room, with a laptop. The person assigned to be the "singer" had to perform the song, while the person assigned to be the "producer" had decision-making power—final say about everything from the genre and lyrics to performance and recording. We video-recorded their brainstorming conversations, including the part where they sang their song hurriedly into a microphone. (This might be an even better part of my job than watching people go on speed dates!)

This task required certain things from each partner for success. The producer had to be unafraid to make concrete decisions, but also had to understand the comfort level and performative strengths of their partner. The singer (even if they couldn't sing at all) had to come from a place of yes—a creative partner who could give the producer the confidence that they could succeed as a team.

When the fifteen minutes were up, the participants gathered in a large room, listened to all the songs from the session, and voted for the best one. The winning pair earned a twenty-dollar payment, and the producer decided how to split the "royalties."

One of my favorite conversations was between John (singer) and Claire (producer). In the very first turn of their chat, Claire asks about John's singing preferences:

CLAIRE: What type of music would you like to sing?
JOHN: Yeah. Umm . . .
CLAIRE: Your voice is deeper, right, so it would—
JOHN: Something soulful, maybe funky.
CLAIRE: All right. I love it.

Claire is getting John's perspective and cheering for him—for their song—from the start, by asking about *his* preference before considering her own. As the conversation proceeds, she uses respectful language. She makes him feel that it is good to be with him, using positive language like "Great idea," "You killed it," "We got this," and "I'd listen to this on the radio." She shows John that he's worthy of care by acknowledging his feelings: "Shake off those nerves" and "Do you think you could do better?" She's listening actively—nodding and smiling—but she's also listening *responsively*, with affirmations like "I love it" and "That was amazing" and excited back-channels like "Yeah yeah yeah yeah yeah." She uses responsive call-backs to great effect. At one point, she admits her weakness as a producer: "Do you want to do like a topic, like, I don't know. I'm terrible at this. I can't write songs." And again: "There's so many things that rhyme with 'free,' but I can't think of anything for the next line."

Claire's vulnerability encourages John to offer a great topic for their song—"Food!"—and to reciprocate when he's struggling, too: "I just can't sing." A few minutes later, Claire calls back to this moment to affirm John: "There you go. Look at you! You *are* musical." She remembered his vulnerable disclosure and called back to it to encourage him: responsive listening in action. By the end, they've written an adorable ditty:

I got a rumble in my tummy.
Need a fill of something yummy.
And you know I like it gluten-free.
I can't have wheat or that flour.
I need nutrition, yeah, that power.
I'll even take a sweet homemade treat.
Give me a lemon pop. Yeah, that lemon pop.
Oh yeah, a lemon pop. A homemade treat.
Give me a lemon pop. Yeah, a lemon pop.
Oooh, a lemon pop. A homemade treat.

They record the song, and they're thrilled. "I think we've got a hit single!" shouts Claire, applauding John's performance. "You killed it! You killed it!"

John is grateful, and relieved: "Thank you. I think I—thank—thank you."

Whether it's about writing songs or not, conversation is surprisingly tricky. The TALK maxims—Topics, Asking, Levity, and Kindness—are a way of remembering simple guiding principles as we glide in and out of the relentless coordination game of conversation. Everything we've covered in the book to this point can be considered as a pathway to kindness—they're tools to help people who sincerely care about others to show it in practice. All the maxims go well when we put others' needs and perspectives first, as often as we can. We can prepare and choose topics they want to talk about. We can ask questions they'll enjoy answering. We can avoid boredom, laugh, and be happy, together. We can speak respectfully and listen responsively.

John and Claire's song won the bonus money in their lab session, but the greatest rewards of good conversation go much deeper than winning contests. We seek to connect, savor, protect, and advance—to thrive in every quadrant of the conversational compass.

On the way out of the recording room, John points to his shoe:

JOHN: In other news, I have a fashion emergency.
CLAIRE: Oh no. What happened?
JOHN: So, the bottom of my shoe came off just as we walked in here. What should I do? Like, I don't know—
CLAIRE: There are definitely like shoe shops that can fix that, but as far as like, getting home—
JOHN: That's what I'm—that's what I'm—
CLAIRE: Did you walk here?

The transcript cuts off as they exit the room. The best way Claire could have helped John is unclear—what did he want and need? I like to imagine that she hurried to the Harvard bookstore, using their twenty-dollar prize to grab a pair of flip-flops for him, while he waited patiently in the hallway outside the lab. Or maybe she let John lean on her as he hippity-hopped down the sidewalk to catch an Uber. On the other hand, helping him at all might have been difficult—we are all constrained by time, energy, and resources. But the kindness John and Claire found during their songwriting chat—the common ground they traveled, the trust they fostered—gives us reason to believe that the ending to their story was a good one. And maybe it was only the beginning.

> THREE KEY TAKEAWAYS FROM CHAPTER 5
>
> ## K Is for Kindness
>
> - **Kindness takes work.** Focus on your partner's needs before your own.
> - **Speak respectfully.** Aim to make others feel seen and known, good to be with, and worthy of care.
> - **Listen responsively.** Put in the effort to listen, and show it with your words.

CHAPTER 5 ½

A Quick Pause

Let's take a quick pause. Or a long one. Whatever you need. Let your shoulders relax. We've covered a lot! We've seen that conversation is a surprisingly complex coordination game, and we've covered all four TALK maxims—prompts that help you play the coordination game more effectively. Topics are the building blocks of conversation, while Asking helps us steer from one topic to the next and explore each one, searching for treasure. Levity helps us find the fun (and steer clear of boredom), while Kindness helps us be the conversationalists—respectful and responsive—that we mean to be.

In Chapters 6 to 8, we'll explore situations that test and stretch the TALK maxims—common situations when the coordination game is even more uncertain. First we'll consider groups, where three or more minds converge, putting our TALK skills to the test. Then we'll see how the maxims help us stave off fear and anger when we don't see eye to eye—in a world rife with disagreements and differences. And finally we'll consider apologies, one of the most powerful tools we have to repair and build our relationships over time.

Grab some water (or a beer or a glass of wine). Take a deep breath. Are you ready?

Let's *go*.

CHAPTER 6

Many Minds

ONE DAY IN MY TALK COURSE, we were discussing work meetings. The students had a lot of thoughts about work meetings—they'd all experienced so many bad ones. We'd gotten pretty far into the weeds about the intricacies of agenda design, the best food and drink to consume in front of other people, and the relative pros and cons of Zoom versus Microsoft Teams. That's when one student, Raj, raised his hand and steered us in a new direction.

"I don't feel that nervous at meetings, but I feel *really* nervous at parties," he admitted. The topic shift was intriguing. "It's usually when other people are talking and I find myself standing alone, for whatever reason." The other students nodded their heads in knowing agreement.

Party panic is a familiar feeling. It happens when you find yourself in a very unstructured group context, standing alone like an awkward island. With so many options, where should you go? Who should you talk to? Should you sidle up to the guacamole, mosey toward the bar, look at your phone—or make a beeline to the door?

Another student, Annie, offered Raj some advice that a friend had given to her. "If you find yourself alone at a party, you can just interrupt any pair of people talking," she said. "They will be relieved that you've joined."

Raj found this advice counterintuitive and uncomfortable. What if he didn't know them and had to awkwardly introduce himself? What if they were vibing, and he disrupted their flow? What if they were discussing something private and important and he stymied their progress?

Never one to miss an opportunity for an experiment, I challenged Raj—and any other students struggling with party panic—to give Annie's strategy a try.

A week later, Raj reported back. The party pace of MBA life had given him ample opportunities to test Annie's theory. To his surprise, her advice had *worked.* Over the weekend, he had wandered up to a duo at a party and said hi but not much else. The pair didn't seem bothered by his intrusion at all. On the contrary—they seemed genuinely happy for him to join. They quickly got him up to speed on what they were talking about (a story about a different party), then continued the conversation.

After this successful first attempt, Raj tried Annie's advice a second time. Same thing: it was great. Several other students reported the same—pairs of talkers seemed almost "energized" when they were joined by a third person. *Why?*

Well, for starters, group conversations can be really fun. That's why we have parties, after all. In the right context, group conversations can be exhilarating, uplifting, and refreshing. In part, that's because groups give everyone more cognitive leeway—more time and space to listen and think. In groups, unlike dyads, you don't *have* to speak at each turn (or at all!) to keep the conversation alive, so you have more time to relax and enjoy the moment. Having more cognitive leeway allows us to better use the tools in our toolbox. In the right environment, it can allow everyone's conversational skills to flourish. By joining another duo at the party, Raj was giving them a break—and a fresh set of topics to discuss. Even when we feel like we're "butting in," we're bringing more possibilities than we know.

Still, Raj's hesitations weren't entirely off base. Pulling up a third chair requires some sensitivity and room reading, and it requires kindness from the others to welcome a new member to the group. In fact, from the moment Raj sidled up, the group's conversation took everything we know about the coordination game of conversation and scrambled it even more.

The Kerfuffle

The dyad is the elementary unit of conversation and the easiest to analyze, which is why the science of conversation has mostly focused on small-scale exchanges *à deux*. But we are constantly involved in conversations with three or more people. Dinner parties. Friends crushed against a buzzing bar. Work meetings. Book clubs. Classroom discussions. Bachelor parties. Sidewalk banter. Family meals. Halftime huddles. Carpool chatter. Band practice. More work meetings. Family holiday celebrations in a Detroit basement. (More on that last one later.)

Group conversations are a different beast from dyadic ones. On the one hand, they are full of possibility. Like Raj joining in at those MBA parties, adding more people to a conversational group can provide a breath of fresh air, opening up a deeper pool of stories, jokes, topics, and information. The electricity that emanates from many minds often trumps the energy that even the most energetic pair can muster.

But despite these rewards, in many ways, groups are a nightmare. As group size increases—from dyads, to triads, to small groups, to groups of six to eight, to very large gatherings—so do the chances that listeners' motives, preferences, decisions, and feedback will conflict. One member might be engaged while another is confused. Some might want to joke while others want to be serious. Responding to one might mean ignoring, offending, or boring another.

The number of potential coordination problems within a group scales exponentially according to the number of people in the group, as does the number of one-to-one connections. Four group members means six unique relationships between them, five members means ten unique connections, eight members means twenty-eight, and so on.* Consequently, their coordination decisions become exponentially more difficult, from topic management and question-asking to levity and kindness. Even the simple question *Who speaks next?* becomes fraught when there are many people to choose from.

In a smooth conversation, speakers neither talk over each other nor have too much silence between turns. Dyadic turn-taking is relatively simple: one person speaks, then the other person speaks. But in a larger conversation, it is less clear who should speak next. Sometimes the current speaker selects the next speaker using explicit cues like questions or subtler cues like eye contact. But who should the speaker pick? The person who has spoken the least, the person who has something relevant to say, the person the speaker likes the most, or someone else? Speakers must make these decisions on the fly, and their choices may affirm or offend others in the group. Meanwhile, speakers-in-waiting may have to jockey with each other for the floor, or they may have a turn thrust upon them with little warning. Compared to dyads, groups have much more cross-talk, as well as more frequent gaps, while people are wondering who should go next.

Moreover, as group size increases, the airtime—the total duration of time available to speak while a group is convened—must be divided among more people. In theory, this should mean that each member speaks less often and for shorter stretches. But groups do not tend to split airtime equitably. Instead, as groups become larger, fewer people tend to claim a larger proportion of the available airtime—which can make quieter members feel like they're being boxed out.

Airtime sharing, turn-taking, topic management, innumerable other coordination problems—the complexities of group conversations

---

* $R = [N \times (N-1)]/2$, where N represents the number of people in the group and R represents the number of unique connections between them.

boggle the mind. For instance, in groups of three or more, there's always an audience. It's nearly impossible to connect with any one member of the group on a deep level while others watch. The larger the audience, the greater the uncertainty about how your words and ideas will land with each group member. There are unspoken norms about what's acceptable and unacceptable—what topics are appropriate or ideal, what should be said on each topic, what personal details should be disclosed or kept private, and what types of language or aspects of delivery suit the group, and the group usually doesn't explicitly discuss what those norms are or what they should be. So more minds require more mind-reading—which leads to more guessing and more risk taking.

These challenges apply to all kinds of group conversations, no matter the context. For instance, it's tempting to think of "work" conversations, like formal meetings, as totally separate from "nonwork" conversations, like casual social gatherings with friends. But most group conversations don't fall neatly into the categories of work or nonwork. Is an office Christmas party a work or a nonwork conversation? What about drinks at the bar with your co-workers? What about at your house? What if your intimidating boss is there? What if he's wearing ripped jeans rather than his normal wool suit? What about a gathering of your college friends who might be able to help you get a new job? What about that moment at a holiday gathering with cousins when the conversation suddenly shifts to a seemingly viable entrepreneurial business idea? Most of our conversations aren't squarely work or nonwork, formal or casual, but something ambiguous in between—and as we've seen, the timbre of any conversation can shift from one moment to the next, careening toward and away from topics fluidly over time.

Secret Hierarchies Everywhere

Among the many things that shift during group conversation is status. Every group has its own *status hierarchy,* an invisible internal ranking of the group members. Social scientists describe status as the respect and prestige we command in the eyes of others. The human mind automatically perceives cues of status, whether or not people are organized by an explicit ranking system (such as the caste system in India, or a formal pecking order in a company). In a military battalion, the group's status hierarchy might be determined not only by its members' formal rank—officer, battalion leader, soldier—but also by each member's tenure in the military, expertise for this specific task or mission, physical stature, and how well liked and respected they are by each of the other group members. Even when there isn't a formal status ranking, there are implicit ones—based on subtle perceptions of qualities like competence, expertise, wealth, attractiveness, influence, or demographic traits like age, race, and gender.* The same dynamics are at play around a family dinner table, where the status hierarchy is influenced by the age and rank of the family members—usually parents wield formal power over their subordinate children, but not always. Here too the status hierarchy is shaped by invisible factors like expertise, relationship closeness, and charisma. And all this rarely gets discussed explicitly, unless someone's dad shouts "Because I outrank you!" or a small child admits to favoring a certain family member: "Today I love Grady the most."

Based on its complex web of explicit and implicit status cues, every group spontaneously develops an implicit status hierarchy. Research suggests that group members perceive it with a surprising

---

* Status (respect and prestige in the eyes of others) is distinct from power (one's control over resources). I focus on status in this chapter, but many of the same points apply to power differences. Status and power tend to correlate with each other, but not perfectly. For instance, most of us know people who have tremendous control over resources (power) but lack respect and prestige in the eyes of others (status), and vice versa.

amount of agreement, even without discussing it. In group conversations, status is working behind the scenes, dictating who says what when, how they say it, what they don't say, and who they're looking at while it all unfolds. And it's the key reason that group conversation is so challenging.

And, maddeningly, status hierarchies aren't static. Think about a seemingly straightforward casual gathering like this one: Five friends meet at a bar during a Thanksgiving holiday. There's a pair of sisters, the boyfriend of one sister, the other sister's work friend, and another guy they went to high school with. The high school friend, Fred, isn't a tightly knit part of the group, but everyone knows Fred's the most financially successful of them, while the sister's boyfriend is the funniest, and the sister's work friend is the most physically attractive.

We wouldn't normally think of status at a bar during Thanksgiving holiday, but it's operating in major ways, even here. The order of their rank isn't talked about, and it's likely to change from one topic to the next as they shift from talking about Elon Musk's new Tesla vehicle, to gossiping about their old high school friend with six kids, to commenting about the Kardashians, to reminiscing about that time they had a massive bonfire in the woods. It's possible for any group member to go from high status to low status to somewhere in the middle, all in the same conversation.

Gaining more awareness of when we have high or low status can help us make better conversational choices from one moment to the next. In the same way that we can think a little more consciously about our conversational goals, we can also think more consciously about the context—namely, the other people in the group, and the changing status hierarchy among them.

Why Claiming Airtime Matters

One of the key ways status impacts group conversation is through the allocation of airtime, the main currency of conversation. While one-on-one conversations make it easier for both parties to participate relatively equally, group conversations make it easier for high-status individuals to dominate and low-status individuals to be pushed to the margins. High-status conversationalists tend to take up more airtime and blurt their thoughts aloud with little hesitation, while lower-status members stew in silence. This fuels a vicious cycle: those who speak more are seen as having higher status, which grants them more leeway to speak, even on topics on which they don't have value to add.

Airtime allocation has significant consequences. If an intern at a marketing firm doesn't get airtime to explain why she thinks the firm's advertisement won't land with Gen Z, the ad may flop. If the funny boyfriend in the bar-hopping group at Thanksgiving stays quiet, the conversation—and in time the group's relationships—may fizzle. If the youngest child never speaks at family meals, he may never learn to speak in groups, and the family can't learn from his creative insights.

Even worse than these informational and relational costs, scientists warn that implicit perceptions of status tend to reproduce discriminatory social hierarchies, like hierarchies of race, sexuality, age, and gender—meaning that voices that are marginalized by society are also marginalized in particular group conversations. If the woman in a room full of men doesn't ask about updating the company's maternity leave policy, will it ever be updated? If the cash-strapped friend doesn't joke about the high price of drinks, will he (and anyone else thinking the same thing) be able to afford the next bar crawl?

Airtime determines what the group learns, believes, and decides, and it also reflects who the group is and what it cares about—what topics the group believes are worth discussing, and which voices it believes are worthy of airtime. In this way, status hierarchies get in the way of good group conversations—and good societies—that accurately represent their members.

Gender: A Case Study of Status

To observe the effects of status hierarchies on group conversation, let's start by looking at the floor. The floor of a legislature, that is. It's where members speak—where they are said to "have the floor." The House of Commons and the House of Lords in the United Kingdom, the U.S. House of Representatives and Senate, and many other governmental bodies around the world have floors with established processes and rules. They're a far cry from the unstructured dyadic conversations we've considered in this book thus far. Their group interactions are highly structured, with rules outlining who is to speak, on which topic, and for how long. The rules—a variety of established practices and ad hoc arrangements—are enforced by presiding officers, who keep the floor in order. These systems, devised centuries ago, ensure that everyone gets to say their piece in an orderly, efficient manner. I bet our Conversation Age dinner party host, Immanuel Kant, would have loved their orderly ways. But, just like Kant's, the rules are not all good.

In 2011, the psychologist Victoria Brescoll, an expert on gender dynamics, studied speaking patterns on the floor of the U.S. Senate. Her analysis suggests that even at this upper echelon of the professional world, female senators speak significantly less than male senators. While male senators with more power spoke more than male senators with less, power didn't influence female speaking patterns—women at all levels of status spoke less. Brescoll showed that this difference was driven by women's concern that being highly outspoken would result in backlash—that they would be perceived more negatively than their equally outspoken male counterparts. And these assumptions were correct: observers (both men and women) viewed outspoken women more negatively than outspoken men.

As you likely suspect, these effects appear not only in the highly regulated realm of the Senate. In all kinds of group conversations, studies have found, women participate more cautiously than men do, in part out of an attempt to avoid backlash. This dynamic can be understood as a consequence of status. In the status hierarchy of many groups, women tend to occupy a lower place than men, particularly in the workplace or any context where women have been historically underrepresented.

Lots of research shows how women are more selective about when to speak—they speak less, even when they know more. For instance, when the economist Katie Coffman asked men and women to rank themselves in a lineup of who should speak, women put themselves lower in rank (farther back in line) on average, whereas men with the same amount of knowledge put themselves closer to the front of the line. Women also tend to prefer interacting in smaller groups of two to four people, which feels more personal, while men prefer larger groups of five or more. And even in relatively small groups of three people, women still tend to speak less than male group members. In any context, one (somewhat dissatisfying) solution is that women can endeavor to speak more. But it would be even better for men, due to their higher average position in status hierarchies, to help.

In working to make group conversation a success, which, as we've seen, is more complicated than making dyadic conversation a success, high-status members (in this case, men) should endeavor to lift up lower-status members (women) and make them feel more invited to speak. Of course, this isn't just about gender. When the junior assistant is struggling at a work meeting, or when we bring a new acquaintance to hang out with a group of old friends, or when no one in a family seems to be listening to the youngest cousin, high-status group members should ask themselves: *What can I do to mitigate the adverse effects of the status hierarchy, the silent puppeteer that's pushing low-status members to the margins?*

Perhaps the most important thing high-status group members can do is to simply get out of the way. Yes: Stop talking! Give others more space to talk! This may seem obvious, but it is trickier than it first appears. It's hard for you to *not* dominate a group conversation when you're used to leading or have expertise, and others seem quiet or lost. But by staying out, you can invite others to speak. Giving others space is an incredible gift, although it can sometimes take relentless effort to achieve.

Still, to really allow for good group conversation amid status differences, high-status group members can't *just* stay out of the way. They need to actively cultivate in their group the elusive phenomenon called *psychological safety*—the belief that you won't be punished or humiliated for speaking up with ideas, questions, concerns, or mistakes. In conversation, you need to feel safe to be who you are and say what you think—to feel like you can express more, if not all, of yourself, without negative repercussions.

To voice their ideas freely, as Amy Edmondson and colleagues have shown, group members on the low end of the status hierarchy require psychological safety—without it, they'll remain inhibited and marginalized. But the more members a group has, the harder it is for low-status members to achieve psychological safety, because the more people there are to judge them. In this way, group conversation presents a paradox: as group size grows, the more the group needs psychological safety to succeed, but the less likely it is to achieve it. In this paradoxical environment, we'll consider what high-status group members can do to foster trust from above, and what low-status group members can do to participate confidently from below.

Fostering Inclusion from Above

When I was in graduate school, I attended a party at a professor's stately home, which had an avant-garde art collection and a surprisingly large private yard. At one point, I needed to escape the comedy of shoptalk and social errors inside the house, so I stepped out into the back. There I found a woman sitting on the steps admiring the garden.

"Hi, I'm Alison. Can I join you?" I asked, already halfway to sitting.

"Please!" she replied.

I asked if she was enjoying her visit, and she said she had recently returned from a sailing trip, and honestly, work travel was making her miss the sea. Having grown up sailing in the Finger Lakes with my dad, I was grateful for this common ground, which offered a clear path to move up the topic pyramid. I asked about her trip and her experience sailing. She was excited to tell me how she and her husband had finally purchased their own boat a few years earlier, and she seemed eager to hear about my experiences boating with my family. She was warm and sharp and funny, and I loved our conversation—I had so much to learn from this pensive sailor.

But then a third person joined us. He was seemingly starstruck to meet my conversation partner, though he too was a professor. "I'm a longtime fan of your work," he gushed.

His praise made it clear that the woman on the back stoop was very well respected in our field, though I hadn't realized it. She engaged with our new group member warmly, laughing about the many mistakes she had made in her work over the years. He peppered her with questions about her research and teaching, while subtly—and perhaps unintentionally—boxing me out of the conversation.

The addition of our third interlocutor had shifted not only the topic but also the emotional tone of the conversation, and I felt invisible. I listened politely for a while, unsure how to get back into the conversation—unsure how to add value to two professors who were much higher status than me. I looked toward the back door, eager to slip back into the party.

That's when the woman looked our third group member in the eyes and asked, "Oh, do you know Alison?" He turned to me as she continued: "She grew up sailing in the Finger Lakes, and she's doing really cool research about anxiety." And just like that, she'd pulled me in. She saved me from my low-status, escapist panic. She made me feel valued. She made me feel seen and known.

This incident was memorable because acts of conversational inclusion from above are relatively rare, not just for me but for most low-status group members. Those with high status tend to use less respectful language and listen less attentively. Research by Adam Galinsky and his coauthors shows that possessing high status inhibits perspective-taking overall. To put it bluntly, having more status makes you less kind. Whether you're a seasoned team leader at work, the best player on your volleyball team, the most popular kid in your class, or principal cellist in the orchestra, you as a high-status group member need to work extra hard at listening and learning about the perspectives of other group members—and pulling them in.

Luckily, high-status group members aren't *always* high status as they move through their lives, which can help. My research with organizational scholar Catarina Fernandes suggests that experiencing different levels of status in different situations makes us better at perspective-taking—sitting on the bench in basketball prepares you to speak respectfully and listen responsively to a lowly intern at work when you're the boss. Being the lowly intern at work can help you relate to a new member of your friend group—or to communicate more empathetically with the confused nonfinancier when the rest of group starts talking about crypto. I suspect the sailor-professor on the back stoop knew how I was feeling because she, too, spent many years navigating a male-dominated field, feeling marginalized and boxed out. We can draw from our own varied experiences of status to understand how others might be thinking and feeling, and we can lean on the TALK maxims to draw them in.

HIGH STATUS AND THE MAXIMS

When high-status members speak, there are many ways they can foster inclusion and trust, starting with acknowledging lower-status group members and giving credit generously. If you call back to a point made by a group member, it's quite meaningful to give them credit for it by name—and to make sure we don't erroneously give credit to dominant group members by default. The human mind tends to automatically attribute good ideas to high-status or dominant group members. But knowing that this glitch in human memory exists, we can seek to correct it—to make sure we give credit to the right people, to make them feel seen and known.

Being vulnerable about your failures can also help low-status members feel more secure. Research by organizational scholars Adam Grant and Constantinos Coutifaris shows that if you're high status (or explicitly a leader), openly discussing criticisms you've received about your own performance—like the sailor-professor when she admitted the mistakes she made as a graduate student—makes other group members feel less afraid to fail and to share their failures with others.

In my own research, I've demonstrated that when high-status group members reveal their failures—among strangers, co-workers, and entrepreneurs—it increases others' admiration for them and reduces others' destructive feelings of interpersonal envy. These effects are particularly strong when high achievers share the struggles they've encountered on their path to success. High-status vulnerability works well in personal conversations, too. Fred, the wealthy friend at the bar at Thanksgiving, could talk about how his first few years on the job were a disaster, or how he isn't sure his work is fulfilling beyond just the money. Veronica, the beautiful high school friend, could joke about her obsession with skin-care products. Vulnerability from above begets feelings of safety for all.

Levity from high-status people in a group conversation can also foster trust. Particularly in serious moments, or when low-status members have grown quiet, they can make small moves to inject self-deprecating humor, make genuine compliments, or shift to topics on which lower-status group members have expertise ("Did you know Alison grew up sailing in the Finger Lakes?" or "Did you know Sheena is a talented statistician?"). Such moves can be a welcome relief and can empower low-status members to share their ideas.

Kindness from high-status group members doesn't always require words. Nonverbal cues can be surprisingly effective, too. Take eye gaze as an example. My research with Nicole Abi-Esber and Ethan Burris studied groups of people during work meetings and found that giving eye gaze conveys attention, listening, and care, while receiving eye gaze, especially from a group leader, empowers people to speak their minds without putting them on the spot. Unlike saying "Angela, what do you think? We haven't heard from you in a while," helping people feel seen and known by using inclusive eye contact can make them feel empowered to speak when they have thoughts to share.

The power of inclusive gaze is especially helpful for members of marginalized demographic groups, like women and ethnic minorities—those who tend to feel the most invisible. Equitable eye gaze from a leader in mixed-race groups confers more equitable perceptions of status among the group members. Looking at everyone means resisting the urge to mostly look at the highest-status members of the group, or to look only at the people talking, an instinctual impulse. Though making more equitable eye contact won't solve systemic discrimination, trying to give eye gaze to everyone, especially low-status group members, is an easy place to start.

A few years ago, on a Home Depot run, my dad and I found ourselves deep in conversation with a couple of contractors and Home Depot employees. Though I'm a competent DIY-er, my dad is a master. He built the house I grew up in, and he has degrees in different types of engineering—chemical, mechanical, electrical . . . I've lost track. We were renovating the basement of our lake cottage, making a bunk room for my kids, and we needed advice about some lumber in one aisle and plumbing supplies in another. I had decidedly low status—the only woman in a group of men in a stereotypically masculine context, with the least amount of knowledge about lumber and plumbing. I wasn't being marginalized in a nasty or dramatic way, but status was at play. The other group members mostly addressed each other, not me.

But these situations are where my dad shines. He makes everyone feel valued, and that includes his daughters in masculine domains. Just when I felt like I couldn't take another moment of technical chatter about gutters, drains, and bathroom fans in which I was subtly excluded, he would throw in a loving jab about my lack of attention to pull me back in, make a self-deprecating quip about his technical obsessions with lightbulbs, or give me a twinkling smile to show we were on the same page. Sometimes he just looked at me, acknowledging that I was there and worthy of inclusion. Even while an unwieldy team was rifling through pipes and seals in the plumbing aisle at Home Depot, he found the fizz and kept everyone involved. We all can.

Participating from the Margins

Cal is a well-respected media executive and self-described extrovert, but even Cal sometimes finds himself in a low-status position. When I talked to him, he told me about a time he felt contorted by the nuances of status. His best buddy, Michael, had invited him to the opening bell at the New York Stock Exchange. Michael was attending because one of his clients was ringing the bell that day—they'd just announced a big partnership with a Fortune 500 company. Once they arrived, Cal found himself in a conversation with Michael and the client. Cal knew Michael really well—they were best friends—so he understood (and cared) that the interaction was important for Michael's work. "I wanted this guy to like me, to have this reflect well on Michael. Plus, I'm a great conversationalist! I can win anyone over!"

But it didn't go great. Cal was a bit underdressed for the occasion, so he felt out of place, and the client insisted on making shoptalk with Michael, which Michael needed to indulge. "Michael was trying his best to include me, because he gets it," Cal told me, "but he was also their guest." Ultimately, he felt "trapped a little outside the conversation. I couldn't quite speak the language, and I felt like I was letting Michael down. Plus I felt like the client just wasn't that interested in engaging with me. It was super frustrating."

As we've seen, status isn't static. It changes not just over our lives and careers, and not just from one conversation to the next, but even from one topic to the next. That's why many of us feel daunted in groups—our rank in the status hierarchy can drop precipitously at any moment. Cal knew he had high status with Michael. But in an unfamiliar context in which he wanted to support his friend, in which he knew nothing about their industry, he had decidedly low status.

Unlike my sailor-professor savior and my dad at Home Depot, Michael's high-profile client didn't pull Cal into the conversation. This is common. In most groups, low-status conversationalists are forced to live with a lack of involvement and trust—embracing the difficult reality that psychological safety may be quite difficult (or impossible) to achieve. This is especially true if the group is quite large and comprised of members who have many conflicting motives, egregious power differences, or who just aren't great at pulling others in.

Conversationalists with low status usually have to figure out how to make the best of a limited array of options. Because they tend to get the smallest share of airtime, they have fewer chances to say what they want to say and accomplish their conversational goals. On top of that, people with low status are severely limited in *what* they feel they can say. Research by the psychologist Adam Galinsky suggests that "status determines your range": group members with low status feel that they have a much narrower range of topics, questions, and jokes that will be seen as appropriate to contribute.

For low-status group members, Galinsky's work suggests that "low power is a credibility crisis." They constantly need to prove their credibility, and even when they do, their insights are less likely to be listened to or taken seriously, like Cal's nascent thoughts about Michael's industry. These dynamics put them in a double bind. If they don't contribute, they may go unnoticed. If they do speak up, they're more likely to get judged negatively for stepping outside the acceptable range of actions.

The double bind leads low-status group members to do odd things. They stay quiet when the group could benefit from their ideas—the group loses their valuable humor, dissent, and feedback. Staying quiet may be a wise move for a newcomer to a group, but if you succumb to silence indefinitely, you may be stuck with low status indefinitely, and the group won't benefit from your insights.

When low-status group members do speak up, they tend to adopt a host of "impression management" behaviors (a Goffman phrase) to compensate for their fear of appearing incompetent or unlikable. They are more likely to use technical jargon, in an attempt to convey their competence and belonging, and to laugh politely, to appear competent and likable. They are less likely to ask for advice or to reveal their struggles and failures. Cal ended up mostly offering encouragement and congratulations at the New York Stock Exchange, but it felt "pretty vapid and sort-of embarrassing," as he explained later. Sometimes we need to embrace a status hierarchy, knowing that it's only for the duration of a certain topic or conversation. Especially here, the TALK maxims are our friends. We can use Topics, Asking, Levity, and Kindness to identify the moves that are both high-payoff and realistically feasible, under circumstances that aren't ideal.

LOW STATUS AND THE MAXIMS

In any conversation, whether in a group or a dyad, the person who chooses to switch to a new topic, or who assertively encourages the conversation to stay on the current topic, or who calls back to a topic they'd left behind is called the *topic leader*. In most informal contexts, like dates, family dinners, and chats around the water cooler, talkers tend to fluidly trade off the role of topic leader, so low-status members can easily jump in with a new topic when there's a lull. Even in more formal contexts, when the topic leader is often dictated by the status hierarchy, people with low status can still practice topic leadership (even when it doesn't seem like it). If you keep your senses tuned to the group's goals, you may start to notice when the group is getting too far off topic. This is an opportunity to jump in.

Acting as momentary topic leader doesn't require much airtime—all it takes is giving the topic a subtle nudge with topic encouragement ("Can you tell me more about why jet fuel is so expensive?") or shift ("Can we get a new round of drinks? What does everyone need?"). Further, paraphrasing—repeating, summarizing, or reformulating what previous speakers have said—is one of the most effective group behaviors available. Low-status group members may be perfectly poised to paraphrase by listening to and summarizing what more vocal group members have said.

Perhaps the most approachable way to contribute to a group is by asking a question. It's a short, self-contained contribution, but it's incredibly powerful. Asking questions doesn't require high status or topic-specific expertise—in fact, ignorance is an asset. The curse of knowledge often prevents expert group members from communicating important details clearly to the whole group—they don't know what others don't know. This happens everywhere, like when two friends who have seen a TV show start talking over third group member who hasn't. In situations like these, when more information is needed, low-status group members are perfectly poised to ask follow-up questions. There is no end to the questions that Cal could have asked Michael's client: How did he land in the industry? What were some stumbling blocks he overcame? Would he stay in that industry forever?

Then, blessedly, there's levity. Lower-status group members who wield humor well can more easily climb the status hierarchy in their groups and organizations. Individuals who make funny and appropriate jokes are more likely to be nominated to leadership positions by their peers—even after just *one* successful joke.

Telling jokes is daunting, but remember (from Chapter 4) that levity moves don't have to be funny. Levity is any fizzy, warm moment that helps the group stay engaged. Give a compliment. Call back to a funny moment. Laugh, or at least smile, freely. These moves don't have to be complex. We all love it when a group member's *only* contribution is a mildly funny quip or compliment or topic shift or giggly laugh that lifts the mood. I like having my colleague Pete in meetings simply because he smiles a lot. Perhaps Cal didn't need to feel so frustrated about offering encouragement and congratulations to Michael's client—levity itself is a meaningful (and kind) contribution.

A Christmas Tradition

Forest-green tablecloths, crystal candlesticks, cream tapered candles—the decor evoked a soothing mood that starkly contrasted with the mayhem happening around it. Tiny flames danced wildly in the wind as people were talking, loudly, often at the same time. It was December 24, and my husband's Polish Catholic family was celebrating Wigilia, the traditional Polish Christmas Eve supper. Derek had proposed to me a few months before, and it was my first Christmas with his family. At Wigilia, children keep an eye out for the first star to appear in the night sky. That's when fasting ends and the feast—pickled herring (*sledzie*) and red cabbage with pork (*kapusta*)—can begin. (Perhaps it's no surprise that Derek and his sisters went through the McDonald's drive-through for milkshakes and french fries on their way to the party.)

His family had been hosting Wigilia feasts in the basements of their homes outside Detroit for decades, since immigrating from Poland in the 1930s. The size of the crowd varies from ten to fifty people from year to year, depending on who comes to town, and who's hosting whom. But they always host the Wigilia feast in the basement—per time-old traditions, and because it's impossible to fit that many people anywhere else.

Having grown up in an Irish Catholic family, I was no stranger to large holiday gatherings, but I'd never seen one quite this large: forty-two people in one room. Loud voices, midwestern accents, lots of interrupting—no lulls. I sometimes couldn't tell if they were mad at each other, joking, or both. And I was quickly learning the rules: a certain aunt always sits at the head of the table; an uncle retreats to the farthest living room so as not to be disturbed; the kids aren't allowed in the kitchen. Secrets were exchanged upstairs, in puffs of hairspray and bronzer. Running and shouting and dancing and laughing pervaded every corner of the house. It was next-level family chaos, teeming with life. Luckily, Derek had warned me in advance, knowing that the dynamics might not jibe with my hope for orderly, attentive conversation; "It might seem a little crazy at first, but don't worry," he told me. "We have a solution."

The cross-talk reached a fever pitch when all forty-two family members sat down at the festive table—a combination of seven folding tables and forty-two folding chairs in the shape of an X. Between ten and twenty conversations were happening at once, jostling and overlapping, as talkers interrupted and laughed in ever-changing groups. At times they were engrossed, and at other times they were uninterested and scanning for other conversation partners. Was anyone actually listening? Surely this melee wasn't sustainable for an entire evening.

But then suddenly, a hush fell over the crowd. Great Busia, the matriarch of the family, stood up, though it was hard to tell. She was just shy of five feet tall, her white hair barely visible over the seated crowd. Still, her voice carried—her presence larger than life. After greeting the crowd, she held up a special piece of bread—the Wigilia *opłatek*—and announced the start of the bread-breaking tradition. Everyone stood, quiet, ready to receive their own piece of bread. Then they shuffled into a new formation: peaceful, orderly pairs. Everyone found a partner standing nearby.

They started talking again, but the chaotic cross-talk had changed.

They broke off tiny pieces of their bread, clutching them between their fingers, as each pair took turns quietly speaking and listening. They reflected about the year that had passed: "I'm missing Ciotka Nica. She was so wonderful." "I've been feeling depressed, but I'm talking to a great therapist." "Congratulations on your new job, Jake." "Your apartment in Vegas sounds great, Ashley." And they shared good wishes for the year ahead. Uncles wished pregnant nieces good luck with their new babies. Cousins hoped for each other's good fortune to get into college, to make it to the state championship, to do fun things together. They told me they were proud of me for going to grad school and wished me well for my wedding next October. We exchanged bits of bread, popped them into our mouths, then shuffled to a new pairing, shoulder to shoulder in the crowded basement. Each pairing was not to exceed two minutes; when the group started to shift, we knew it was time to move.

To my surprise, many people in the room started to cry, eyes misty, cheeks wet. My mother-in-law wept and laughed in equal measure. Later that night she explained her feelings to me: "It's nostalgic, not sad. It's for being reflective, especially if there's loss. You get used to it. You know it's coming, and it forces you to get out of your shell and connect." She met my eyes: "You have to appreciate everyone and everything." For her, the bread-breaking tradition brought a sense of relief—another year had passed with everyone safe. I understood her tears. It's a beautiful tradition.

At the end of the ritual, after all the pairs had met and the bread had disappeared, the group sat back down at the X-shaped dinner table. Left to its natural ways, the gigantic crowd resumed its wild rumpus.

The Steward

It's a wonder the Wigilia didn't devolve into utter chaos and confusion. In fact, it might have—if it hadn't been for Great Busia. She took charge at just the right moment, pausing the pandemonium by asking everyone to listen while she welcomed us to the table. Then she delivered instructions for the bread-breaking tradition, steering us into our rotating dyads. The stories we shared, the wishes we gave, the sentimental tears we shed wouldn't have happened without her deliberate guidance.

To borrow a concept from Elise Keith, the founder and CEO of Lucid Meetings, we can think of Great Busia as a conversational steward. *Conversational stewardship* means supervising or taking care of a group. It's the steward's job to steer the group through the chaos and to help it avoid coordination problems and the challenges presented by status hierarchies. We can all be good conversational stewards in the ways we've seen above, but sometimes someone needs to take charge a little more formally. A good steward plans for differences among group members and gives forethought to the group's goals, topics, and the conversation's structure. Good stewardship is palpable—and so is the lack of it.

While the stewardship mindset is especially useful for people who are explicitly leading a group conversation or hosting a party, we can all benefit from adopting it. The keys to employing it are thinking ahead (*What are our goals? What do people care about? What should we talk about? How could things go wrong?*) and paying attention to the timbre as the group conversation unfolds (*Is the conversation getting to a bad place? Are people feeling good or uneasy?*). The following section discusses tools that stewards can rely on to help the group.

Adding Structure

Many nonhuman species spend a huge amount of time in large groups—herds, gaggles, cackles, mobs, pods, harems, clans. Starlings, medium-size passerine birds with metallic black plumage, gather on telephone wires and in trees. But their most majestic gathering is a *murmuration*—an undulating cloud of thousands that twists and turns mysteriously in the sky. These mesmerizing profusions serve many purposes—predators like peregrine falcons find it difficult to target one starling in the middle of a hypnotizing flock; the group helps each bird keep warm at night; and the group can quickly share information about promising feeding areas. We know a lot about starlings, but it remains a mystery how they fly in murmurations without colliding.

That kind of intuitive, large-scale coordination is beyond the human mind. Get a thousand people together on a football field and tell them to walk in a murmuration, and you'd get utter chaos. The same thing happens in conversations. Sustaining a free-form conversation—with optimal topic management, smooth turn-taking, equitable airtime-sharing, and so on—is nearly impossible in a large group. Indeed, research suggests that as groups scale to six or more people, a single conversation naturally becomes unstable, poised at any moment to splinter into multiple conversations with people talking over each other. Perhaps that's why Immanuel Kant capped the number of invitees at his dinner parties—he knew larger groups are unstable for mutually rewarding conversation.

Here conversational stewards (like Kant) can help by deliberately *structuring* a group conversation. Just as we need traffic laws on the road, sometimes we need explicit rules to guide us through group conversation. While dyads can generally handle unstructured conversation, and three-to-six-person groups adopt natural (albeit dissatisfying) patterns of turn-taking and airtime-sharing, larger groups need guardrails. Even the physical limits can be challenging—so many chairs, shoulders, and knees. But such groups can be structured by creating rules for turn-taking, airtime-sharing, topic management, and so on. These rules generally help simplify coordination problems—making it clearer who talks next and when, what everyone's talking about, and what the goals of the conversation are. For instance, the Wigilia Christmas ritual—passed down for generations and facilitated by Great Busia—added structure to our chaotic forty-two-person conversation. The ritual provided a template for how the night would go, guided the group through what would otherwise be an overlapping mess, and helped people connect in a way they wouldn't have been able to otherwise.

There are two key ways to structure a large group conversation: partitioning and centralizing.

Partitioning. Partitioning is dividing a group into smaller subgroups who talk among themselves. This tool solves many coordination problems because people share airtime structurally rather than through their haphazard choices. It simplifies the coordination game, making the large group more like a dyad.

Partitioning can happen naturally in a large group, as in the informal milling that happened all over the house before the guests settled down for the Christmas meal—individuals moved between small group conversations as they pleased. But it can also be implemented deliberately by a leader or steward, as in the case of the Wigilia breadbreaking ritual. Great Busia led the family through a partitioned conversation: pairs of partners exchanged well wishes along with a piece of bread, and after two minutes they rotated. This could also work with the five friends at the bar during the Thanksgiving holiday or any other casual group. To partition a group, you can just start talking to one or two people and pull others in as needed, or introduce people to each other, drop a topic, and walk away. Groups are remarkably good at adjusting to partitioned conversation, no matter how it starts.

Centralizing. The opposite of partitioning is centralizing, when the attention of all group members focuses on one shared conversational thread. Centralized group conversations often proceed formally by allocating turns through hand-raising, speaking in a predetermined sequence, deferring to an agenda, or giving a chairperson control of the floor. Though centralized conversation loses many of the magical features of private conversation—like improvisation, agency, and serendipity—it allows for a sense of togetherness and the dissemination of shared information to many people. It simplifies the coordination game in a different way.

Centralizing and partitioning are two essential tools for the group's steward, who can think about which arrangement—centralized or partitioned—aligns with the goals of the group and its members, and how long any one arrangement can last before it loses its luster. How we structure group conversations—or not—will profoundly influence what information gets exchanged, who exchanges it, and how everybody feels while the exchange is happening.

As Elise Keith, the CEO of Lucid Meetings, points out, group conversation doesn't have to be centralized *or* partitioned—it can be both, at different times.* Sometimes a group has partitioned, like the friend group at the Thanksgiving bar that broke into two smaller groups, one talking about a new skate park and the other about an old friend who was having a baby; then one person who wanted to tell everyone a story pulled them back together.

Over time, a group may repartition and recentralize many times. Mixing up the formats helps keep either structure from getting old. And they can get old. Centralized conversations can easily deteriorate into dry exchanges between dominant group members as silent observers writhe in their quietude. Such exchanges become uninteresting, frustrating, or even offensive to the silent observers, who may feel uninvolved or invisible. Then it's time to partition. But partitioned conversations can make us feel separated—pigeonholed in an unwanted sidebar conversation and missing out on the benefits of learning from everyone in the larger group. The question for stewards is, what does the group need now?

---

* Keith recommends that meeting managers combine breakout groups and centralized conversation in a sequence that builds in size over time, like 1-2-4-all, 1-2-all, or 1-3-all. In 1-2-4-all, the manager gives some sort of topic or prompt to the whole group, first to consider alone (1), then to discuss in pairs (2), then to discuss in joined pairs (4), and finally to talk about when everyone regathers in the plenary group (all). This progression allows people to think critically on their own, takes advantage of the privacy of dyadic chat, and then relies on the aggregation of knowledge that occurs in larger groups. Whatever the sequence, partitioned and centralized conversations can unfold quickly (over the course of a few minutes) or over long periods of time (over the course of days or even weeks).

Tighten Up, Loosen Up

In terms of stewardship, group conversations fall on a spectrum from tight to loose. If the steward steers the group toward topics they're not interested in while leaving juicy topics unexplored on the table, the conversation may feel too tight. I suspect Kant's dinner guests may have felt overmanaged by his many explicit dinner rules and forbidden topics (like any disagreement with the host about the French Revolution!). On the other hand, when a group conversation starts to feel wild and chaotic—like the unruly pub conversations Kant finally escaped in old age—it's probably become too loose. The steward's job is to find the right balance, sometimes by tightening structure and sometimes by loosening it.

It's easy to imagine deliberately manipulating the structure of work conversations, where agendas, breakout groups, centralized presentations, and icebreakers are par for the course. But the same principles apply outside the boardroom, too. While we may be reluctant to impose structures on our friends at dinner parties or while hanging out on a sidewalk with a few strangers during a smoke break (where the goals are fuzzier, the topics are more personal, and we put a premium on feelings of "naturalness"), a lack of structure can be surprisingly stressful—and awkward—anywhere.

When I attended a bachelorette party in Nashville for a friend of a friend, I was a tangential guest—not in charge. Like most hen parties for women in their twenties, it was supposed to be fun and freewheeling. But, like many conversations, the gathering was insufficiently structured, with fifteen attendees unsure where to go, when to move, and what to talk about. Subgroups were threatening to break off for their own endeavors, everyone was confused and asking what was happening next, and I worried that we wouldn't accomplish the goal of "group fun" that the weekend had promised to deliver.

I've attended other girls' weekends where everyone teases the seemingly stringent planner for devising an itinerary. Who wants to plan activities that should feel footloose and fancy-free? Though we tease them, I've come to deeply appreciate the stewards who take charge, for saying things like "Everyone pick out your favorite feather boa now—we'll need them later for dancing" and "In a half hour, we'll get on the bus to dinner. Our reservation is at eight." Their moves can be small: they launch a planned topic, divide people into smaller groups to play a party game, plan some key entertainment to focus the group's attention, or ask someone to give a toast that centralizes toward joyful celebration. Though directing the group can seem rigid in theory, in practice larger groups need stewards and structure.

Avoid Group Conversation. Just Kidding. Kinda.

These days, when I join Derek's family for Christmas Eve, I relish the pandemonium of the large group, and I also take comfort in the one-on-one connections during the Wigilia ritual. I've learned that when group conversations don't seem to go particularly well, we don't need to feel guilty, overwhelmed, uncomfortable, or frustrated. It's normal! By its very nature, group coordination is chaos. Whether we're crushed against a buzzy bar at Thanksgiving, gathering around a dinner table, preparing for battle, conducting a board meeting, or mingling at a party, we can improve our interactions by becoming more cognizant of status dynamics and maintaining a mindset of stewardship. But bringing together many minds will always be a kerfuffle. And that's okay.

THREE KEY TAKEAWAYS FROM CHAPTER 6

Many Minds

- **Conversation in groups is categorically different** from and more complicated than conversation in dyads. (Don't feel bad if it feels chaotic.)
- Be aware of the **status hierarchy** in a group, which can change, even from one topic to the next.
- Foster a **stewardship** mindset.

CHAPTER 7

Difficult Moments

AS THE SEMESTER PROCEEDS in my TALK class, we learn about the conversational compass, develop inside jokes, practice all the TALK maxims, and spend hours listening attentively to each other in groups large and small. Then, toward the end of the semester, I ask the students to complete an assignment at home. They must decide where they stand on each of fifteen hot-button issues, such as legalized marijuana, taxes on the rich, compliments to someone's physical appearance at work, universal healthcare, genetically modified foods, self-driving cars, gender-neutral bathrooms, gun control, and physician-assisted euthanasia. The students rank their stance on each issue on a scale from 1 to 7, "strongly disagree" to "strongly agree."

I also ask them to come up with three of their own hot takes—"topics you feel passionate about that might be considered controversial or polarizing, or that you think some of your classmates probably disagree with you about." In past semesters, they have mentioned topics like whether transgender athletes should be allowed to participate in gendered sports and whether employees should be fired based on what they've said on social media. I tell them to come to class prepared to discuss their most controversial positions with one another.

You might think: *This is terrifying.* That's how most of my students feel, too. Their anxiety is palpable. When they come to the next class, required worksheets in hand, they're quiet and serious, filled with dread.

I try to manage their trepidation. The instructions on the worksheet I gave them acknowledge that the topics are intense: "Don't be afraid! The topics I've provided are based on the ones used in the academic research behind the exercise, and you can come up with your own topics, too." I tell them that our goal as excellent conversationalists will be to engage on these topics productively, thoughtfully, and perhaps even delightfully. We've waited until the end of the semester for this exercise, and they've been practicing the TALK maxims for several months, so they know that they can pursue multiple goals simultaneously—like learning, persuasion, and enjoyment—even while they navigate charged topics. I assure them that rising to the challenge will be worth it.

To start, I divide them into groups of three. They find their assigned groups and settle into rolling chairs tucked between soundproof panels, so each group has a barrier for privacy. If the weather's nice, they can find a peaceful bench or a patch of grass outside. They'll have a total of three conversations, each one structured as a dyad, while the third member of the group is a silent observer. For each conversational round, I ask the pairs to find the issue on which their ratings are *farthest* apart—a topic on which they have convictions and sincerely disagree. Then they dive headlong into topics they normally avoid.

Difficult Conversations

For many of the students, and for many of us, this is a nightmare scenario. We often avoid disagreements with family members, friends, and colleagues alike. When we think about "difficult conversations," what we're usually thinking about is disagreement. A spat at a family gathering. An argument with your spouse. A salary negotiation. Firing someone. An encounter with a political rival, frenemy, or teammate. We imagine intense emotions—tears, yelling, negative judgments, hurt feelings—and we recoil at the first signs that our convictions might diverge. The very words "We need to have a difficult conversation" can make us shudder.

Some disagreements are smoothed over easily, but others can get out of hand, and you feel the conversation, and maybe even the relationship, slipping away. A disagreement creates a crack in our shared reality—one that can break open into a dark, gaping, angry crevasse. When that happens, disagreements cause us to experience a range of feelings characterized by high levels of both arousal and displeasure.

Recall the chart of emotions—or the "wheel of feelings," as my students call it—that we encountered in Chapter 4.* In that chapter, our challenge was to use fizzy moments of levity to jump out of the bored, disengaged lower-left quadrant. Now we're going to focus on getting out of the upper-left quadrant, where anxiety, stress, fear, anger, and hostility roil (see the figure on page 182). While the lower-left quadrant kills conversation with a silent, poisonous sleeping pill, the upper-left quadrant kills conversation at knifepoint.

The more we find ourselves in that upper-left place, the harder it will be for us to do any of the things that make conversation good. If we're too afraid, angry, or distressed, we can't focus on keeping the delicate coordination game alive. And the stakes of keeping the game alive are high. Disagreements can hurt us tremendously, both personally and professionally, and cause irreparable rifts with family, friends, and colleagues. That means that we must try to manage our disagreements and mend the cracks in our shared reality before they send us spiraling into the upper-left quadrant and beyond. Before our words

---

* Scholars call it the "affective circumplex," because *affect* means "emotion," but no one really wants to call it the "emotional circumplex," right?

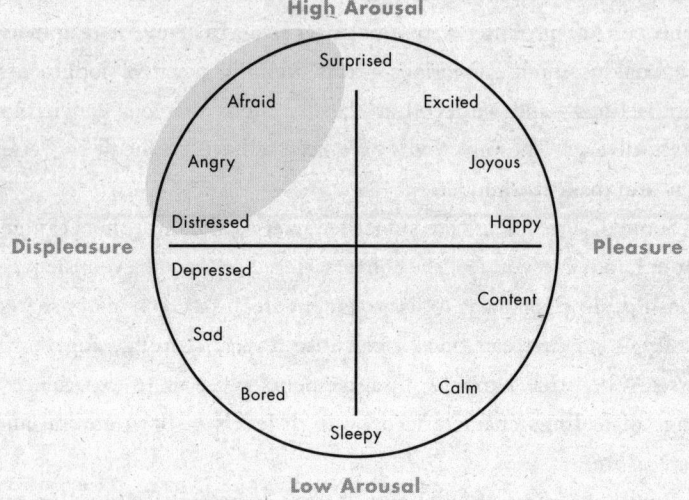

cause harm. Before our rifts become irreparable. Only then can we keep playing the coordination game at all.

Outright disagreements aren't the only kind of difficult conversation, nor are they the only reason we might find ourselves in the upper-left quadrant. Whenever the participants in a conversation have a *difference,* they are likely to have difficult moments. I asked my students to discuss a topic on which their views differ, but many other kinds of differences underpin and swirl around our views—different knowledge, different understandings, different preferences, different intentions, different emotional states, different values, different aspects of our identities, and so on.

In fact, all conversations are characterized by some kind of difference, big or small, because at the very least, everyone is unfathomably different from everyone else in both obvious and nuanced ways (even identical twins like Sarah and me). And every conversation, even "easy" chats that are devoid of outright disagreement, have moments of difficulty: fleeting hints of awkwardness, offense, ambiguity, misalignment, and misunderstanding. In practice, many of these difficult moments are resolved or simply pass quickly without escalation toward the upper-left quadrant. But if mismanaged, difficult moments—even the minor ones—can lead to anger, anxiety, fear, confusion, or competitive arousal. We often can't make perfect sense of those feelings while the conversation unfolds, but we usually sense that *this isn't going great right now.*

A Taxonomy of Difficulty

It can help to think of the far-ranging sources of difficulty in terms of depth, like layers of the earth (see the figure on page 185). Above the surface are *differences in content:* our words, sounds, and nonverbal cues are like trees and mountains and buildings and airplanes—the stuff we can see and hear, in plain view. This is where a lot of coordination challenges live: you say *to-may-to,* I say *to-mah-to*; you wink at me, but I don't see it; I put on a British accent to deliver a joke, you think it's Australian; you laugh at something and I don't, or you laugh at the literal meaning of the joke, and I laugh at the irony of it; and so on. These incongruities may be harmless (or someone skilled can use them for levity), but as we've seen, they can also cause and reflect problems in shared understanding and connection.

Then, at the earth's crust—just above it, slithering in the grass, or just below it, hiding in the soil—we encounter *differences in feelings.* You're thrilled to talk about strategies of war, and I'm bored out of my mind. He lives for political banter, but you can't stand it. She's dying to hear about your kids, but you find her questions intrusive and annoying. You're happy and excited about something happening later that day, but your conversation partner is feeling sad about something that happened that morning. And so on. Incongruent emotions can be ephemeral and changeable, untethered to anything particularly important. Or they may indicate our differing preferences about where to devote our time and attention—signals of something deeper.

When we drill down a bit deeper into the earth, we hit bedrock: *differences in motives*. This is where aspects of our conversational compasses—our goals—are incompatible. He's dying for your advice, but you don't want to give it. He wants to relax, but she wants him to ask how her day went. You want to learn about your partner, but your partner doesn't want to share. I want to figure out my counterpart's best and final offer so I can sell my house for the highest price possible, but they don't want me to know their budget because they want my house for the lowest possible price. And so on.

As we saw with the conversational compass, differences in motives—known, presumed, and unknown—cause all sorts of tension and conflict. As in the Prisoner's Dilemma (which we encountered in Chapter 1), conflicted motives place us in a noncooperative coordination environment. Those motives, like the emotions they give rise to, may be weakly held and changeable, or they may be cemented into their current position after years and years of calcification. This is where some of our big disagreements live—when we seem to be on different tectonic plates entirely. A wife has a great job in Dallas, but her husband wants to move closer to his family in Seattle. An employee wants to keep his job, but the boss thinks it's better if they part ways. You want to put your mom in a nursing home, but she doesn't want to go there, and your sibling can't imagine forcing her to leave her house. How do we navigate conversations around these deeply incompatible motives? How do we use our time together in conversation to make progress? Can we sustain good conversation when a deep, subterranean clash is lying beneath the surface?

Go deeper still into the earth, and you reach the core: *differences in identity*—gender, sex, sexuality, race, ethnicity, religion, age, expertise, personality, physical attributes, and so on. Some identity differences, like race, may be on public display, while others, like deeply held personal beliefs, can be invisible, bubbling deep underground, out of sight—sometimes even beneath our own consciousness.

Unlike the engineering challenge of drilling to the earth's core, engaging in conversation has a way of shooting down to the hot magma of our identities in sudden and unexpected ways. Many of the difficulties we encounter during conversation that *seem* to stem from differences in feelings or motives—closer to the surface—actually stem from tiny fissures that have burrowed all the way down to our identities. The difficulties can emerge even on normal, everyday topics: talking about holidays can strain people of different religions; talking about hairstyles, fashion, music, real estate, or job interviews can feel strange between Black and white friends. We may not be sure if it's okay to talk about our children with childless colleagues at work. Anytime a conversation touches on—or *might* touch on—something about *who we are,* we risk failing to make our partner feel **seen and known** (a violation of the Kindness maxim), and we can easily become defensive, anxious, threatened, or hurt. The molten

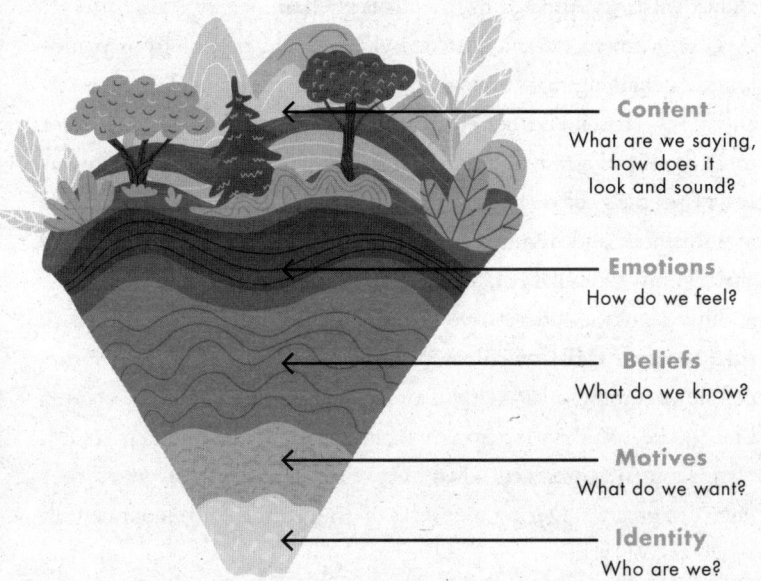

rock at our core can so easily start to bubble, ready to erupt—or cool and harden.

My class exercise—purposefully discussing a topic on which two people have strongly held, opposing views—is designed for the students to practice confronting differences at every level: to aim for clear content above the surface, to be conscious of everyone's emotions at the crust, to understand the differing bedrock of their motives, and to learn and respect each other's core identities. It's their task to keep the conversation going until they can figure out where their differences really lie: At what layer are the rifts? Only then can they reconcile, shift, or accept those differences.

Your Brain on Disagreement

While some people find disagreement exciting, most people, including my students, dislike it, and for good reason. It can be unpleasant and exhausting. And again, the stakes are high.

Disagreement doesn't just feel challenging; it actually is neurologically challenging—biologically taxing to the brain. The neuroscientist Joy Hirsch and her coauthors have shown that "it takes a lot more brain real estate to disagree than to agree." The acoustic properties of people's conversations, they found, differed between moments of agreement and disagreement. During disagreement, the pitch of their voices, the syllable rate, and the acoustic energy were elevated—audible signs of competitive arousal. When participants talked to each other in fMRI machines (the spaceship-like pods that capture people's neurological activity in an array of pixelated, rainbow-colored blobs), the social and visual attention areas in the speakers' brains were more synchronized when they were agreeing than when they were disagreeing. During moments of disagreement, even our neurology is out of sync.

Maxims to the Rescue

With some partners, no amount of conversational acumen will bridge their differences, especially in only one conversation. Deeply held personal values and many aspects of our identities don't often change on a dime. But even tectonic plates can shift. The key is that they shift slowly. For deeply held beliefs, especially those linked to our identities, the only way you might be even a little persuasive is if your partner feels safe with you, and heard by you, so that they remain open, over long periods of time, to the gentle pressure of your differing views.

So what can we expect to achieve within only one conversation? A lot! Every conversation, especially when we stay away from the dreaded upper-left quadrant, can make either baby steps or giant leaps toward identifying, understanding, and bridging our differences. The write-ups from my students show how much can be accomplished in just one good chat. Here's a write-up from one student, describing her conversation:

> I was surprised by how much I enjoyed my conversation with [my classmate] on gun control—a topic I had very staunch views on. In the past, I've entered conversations about gun control with the goal to persuade others that the US should implement severe gun control measures (even going so far as to assert we should ban guns for civilian use). Often, I left these conversations feeling deflated when the other party did not come around to sharing my point of view; for me, the implications (avoiding unnecessary deaths and tragedy) were so evident. During this conversation, though, I found myself referring back to areas of agreement (that we needed some measures of additional gun control) and bringing up the limitations of my view (implementation gaps, skewed cultural perceptions coming from an East Asian country that has strict gun control). I

made a genuine effort to listen to his point of view and he did the same; I appreciated many of the thought-provoking questions he asked me (ex: Is it desirable for the state to have a monopoly on guns, given current distrust of police?). Both of us came away from the conversation with more nuanced views on the subject. And we actually managed to persuade each other on a few aspects without even meaning to.

Using the TALK maxims can help keep even the most serious disagreements from spiraling into the upper-left quadrant. It doesn't mean that some of these conversations won't be painful, but deft topic management, excellent questions, appropriate acts of levity, and heaping doses of conversational kindness can prevent an implosion. If we can avoid the worst ruptures, then we'll always stay in a position to move back toward a rewarding shared reality—toward steady ground.

It's not always obvious how to apply each maxim when things threaten to get out of hand—when you and your partner are having a tense disagreement, when you're not sure if you've hurt their feelings, when your amygdala is firing, or when you and your partner seem separated by a gulf of difference. Luckily, new research provides guidance on how to apply the TALK maxims in these tricky situations. This research has produced strategies for handling disagreements with grace, engaging with identity differences with confidence, and turning down the temperature on a conversation that's getting overheated.

Before I studied conversation, the disagreement exercise I make my students do would have been horrifying for me, too. I've long identified as "conflict-averse," going to extreme lengths to run away from confrontation. But now I know I can do it. And I know you can do it, too.

The Receptiveness Recipe

What can we do to have a successful disagreement, one that avoids the upper-left quadrant? How do we pick apart the specific, concrete features that make a disagreement go well? How can we ensure that these interactions are enjoyable or, at the very least, that they don't become dreadful and hostile? Can everyone leave a conversation feeling better, not worse, when they deeply believe, feel, or desire incompatible things?

Over the last decade, social scientists Julia Minson, Mike Yeomans, and Hanne Collins have been studying these questions. Across hundreds of thousands of conversations, among Twitter users, among Wikipedia editors, and in face-to-face sparring matches, they've parsed the precise language that people use while they disagree. They've differentiated the words that achieve civility and progress from the words that lead to unproductive endings marked by hostility, lack of agreement or persuasion, or irreparable rifts. Their groundbreaking research has unlocked the language of civil (and uncivil) disagreement—and what the best conversationalists tend to say in difficult moments.

Good disagreements, they discovered, are characterized by *conversational receptiveness*—engaging openly with opposing views amid live conversation. Engaging with people of opposing views presents threats at every level of difference, but those who use receptive language are less likely to end their conversation in a hostile blow-up, *and* their partner is more likely to agree with their viewpoint by the end. What's more, these researchers captured a "receptiveness recipe," the linguistic tactics that great communicators (like Primo Levi's Super Greek) use when they're making friends across enemy lines.

In one study, the research team gave hundreds of U.S. citizens a series of controversial topics to discuss, as I do with my students. Here's one example: "Has the public reaction to recent confrontations between police and minority crime suspects been overblown?" After reading this conversation starter, each participant received a message from a conversation partner—one of hundreds of real messages picked by the researchers to present a view that differed markedly from the participant's. A participant who believed the public outcry was overblown—that the 2020 BLM protests were an overreaction—would have received a message like this one, which, contrary to their own view, supported the BLM protests:

> The public reaction has not been overblown; if anything, it is severely muted. Recent confrontations are simply making visible issues that have existed for a very long time. Lynchings and killings of minorities have happened throughout American history. However, now we have proof in real time through cellphone videos and Facebook Live. There are real issues between police and the communities they serve that are only being addressed because of the attention being paid through Black Lives Matter and other protests.

How might the participant—or anyone who was less supportive of the BLM protests—respond to this message in conversation? Due to the complexity of language, the options are nearly infinite. Here are two real responses from participants who were more skeptical of the BLM protests (compared to the person who wrote the first message):

PARTICIPANT 1: Good point. I understand what you are saying. There probably is some truth to the fact that these issues have been hidden for a long time. However, coming from St. Louis and witnessing the Ferguson riots, I can also see how things can be blown out of proportion and make people feel that it is worse than it is. I agree real problems exist, but possibly sometimes attention is drawn in the wrong places.

PARTICIPANT 2: Overreacting to police confrontations can be deadly to the public in general. When animosity toward the police rises, as it has in Chicago, police do not feel safe going into the ghetto neighborhoods. Therefore, those people, in those neighborhoods, literally, have to fend for themselves, because if they need the police and call for their help, the police can't help those in need there, because they will likely be shot at themselves.

If these responses feel different to you, you're right—they are different. When the research team recruited third-party observers to rate these two statements in terms of receptiveness, the ratings they got differed dramatically. The observers considered the response of Participant 1 to be one of the most receptive in the whole dataset, and the second to be one of the least receptive. You can likely sense that the first response was the more productive approach to disagreement. But why?

Let's unpack the ingredients of the receptiveness recipe:

- **Acknowledgment.** (*"I understand what you are saying . . . these issues have been hidden for a long time."*) Straightforward as it seems, simply repeating what your partner has said brings many benefits. First, it makes your partner *feel* heard, a key piece of responsive listening that becomes even more important amid disagreement, when emotions run hot. Acknowledgment also brings informational rewards—it helps us identify areas of misunderstanding so that they don't escalate into bigger problems. And it brings memory rewards, helping us remember what our partner said. Repetition makes things more memorable. Repetition makes things more memorable.
- **Affirmation.** (*"Good point."*) Affirmation is acknowledgment's more effusive cousin. You not only restate what's

been said, but you also attach a positive evaluation to it. You like it. You love it. You can't live without it. Everyone craves validation, yet we often forget to give it. My fellow researcher (and stand-up comedian), the psychologist Adam Mastroianni, taught me that although "Yes, and" is the first rule of improv comedy, the second, lesser-known rule is "Treat your scene partner like a genius." Don't just grit your teeth and agree to whatever your partner says because it's improv and you have to. Actually fall in love with their choices and their mind, however weird, unexpected, or oppositional—and show it. What this means in the context of conversation is . . .

- **Flagging points of agreement amid disagreement.** (*"I agree real problems exist."*) When people disagree, they tend to fixate on the issue of disagreement, omitting (and forgetting) the millions of other things in the universe that they agree about—ice cream, gentle ocean waves, music, good books, cozy blankets, twinkling lights. Whatever the topic, there will always be aspects of it on which we agree, and aspects on which we disagree. Further, disagreeing at all on a particular topic reflects implicit *agreement* that we believe the topic is worthy of airtime.

  Alas, we often forget to say these points of agreement aloud. But making small references to sources of agreement can go far. It can help you seem like a reasonable person, and both of you can continue to feel connected, even as you disconnect on other points. The psychologist Roderick Swaab recently found that it's important to identify the goals everyone wants to achieve *before* they dive into intense negotiations. Negotiators who find points of agreement go on to treat each other more congenially and create more value overall even as they disagree. The receptiveness recipe reminds us that we can sprinkle points of agreement into our conversation *while* we disagree—and afterward—as well.

- **Hedging your claims.** (*"Probably is some truth . . . possibly sometimes attention is drawn in the wrong places."*) Amid disagreement, hedges like "maybe," "possibly," "I wonder," and "it's likely" qualify your claims and help you seem reasonable, and capable of considering nuance and complexity. The speaker knows that no single motive is universally good, no single emotion is necessarily virtuous, no identity is superior to another, and no position on any issue can be known (or proven) with certainty. As a scientist, I have come to appreciate this deeply. In pursuing the truth, you are unlikely to accumulate enough evidence to be 100 percent sure (nor sufficiently nuanced in your understanding of the truth's caveats, exceptions, and boundary conditions).

  Hedges help us humbly acknowledge our own uncertainty on the fly. You might remember that hedging was rated as highly respectful back in the police stop study in Chapter 5, too. For most, using hedge-type language feels counterintuitive because our instincts tell us to be resolute, strong, and decisive—especially if we want to persuade other people that we're smart, that we know the truth, and they should believe us. But research on receptiveness suggests that conversation partners are actually more likely to believe and listen to people who seem reasonable, balanced, and open-minded.

- **Positive framing.** (*"The police can't help those in need because they will likely be shot at themselves"* is negative framing. Positive framing would be: *"Police officers seek to protect citizens' safety as well as their own."*) Just as disagreeing takes more cognitive effort than agreeing, our brains more easily receive arguments framed with positive language than those framed with negative language. Again, this corroborates the findings of the police stop study. Our minds are naturally averse to loss—we're more dramatically triggered by the prospect of loss than by the prospect of gain. In conversation, this means we should try to focus on the benefits we might achieve rather than the costs we might incur. Positive framing is easier for our partner to process and engage with, and as we've seen, it simply makes them feel that we find it good to be with them.

- **Sharing personal stories.** (*"Coming from St. Louis and witnessing the Ferguson riots, I can also see how things can be blown out of proportion."*) Sharing your personal history—especially experiences involving vulnerability or being harmed—helps others understand your identity and why you may have come to hold a certain motive or opinion. Sharing personal stories is humanizing, helps us connect, and fosters mutual respect. In contrast, reciting explanations or facts you've learned, which can be compelling in a written report, feels argumentative in live conversation.

- **Avoiding "explanation words."** (*"Therefore, those people, in those neighborhoods, literally, have to fend for themselves, because if they need the police and call for their help, the police can't help those in need there, because they will likely be shot at themselves."*) Explanation words like *because, therefore, always,* and *never* can make us sound dogmatic, pedantic, and condescending. Though explanation words might work in one-way communication (like public speeches) or written text, most people don't like being lectured to during an interactive dialogue.

Luckily, my students were more prepared to handle intense disagreement than many of the study participants were. By the time they embarked on the hot-button disagreement exercise in class, my students had learned about the receptiveness recipe, and they had practiced using it. They'd also learned to think of receptiveness as a stress test of their kindness: they could listen responsively and use respectful language *even when* they were talking with someone who disagreed with them. While respectful language and responsive listening are a terrific start, moments of disagreement often require the specific guidelines of the receptiveness recipe.

In class, each pair of partners talked about their hot-button issue for ten minutes. As they dug in, I walked around to listen, moving quietly without hovering anywhere for too long. I could hear them using the receptiveness recipe—hedging, acknowledging, and affirming each other. It was like watching an ice cube melt. I didn't hear any raised voices or see any tears, but I did notice sustained eye contact, nodding, some smiles, and believe it or not, *laughter.*

After their conversations, we regathered as a whole class. It was hard to get them to stop talking. Once they settled back in, I asked them to describe their experience in one word—to just shout what came to mind: "Amazing," "So fun," "Eye-opening," "Best thing I've ever done," "I wish they taught this in the first year of the program," "I wish I learned this in high school." These weren't the voices of a positive but vocal minority. My students list this exercise as the most popular session in the course every time I've taught it. Around 40 percent of the students list it as their favorite, a whoppingly high percentage, given that we do more than twenty exercises during the semester (exercises that include calling their best friend, and a humor workshop!). When people are armed with the receptiveness toolkit, they can sort out even the thorniest differences. This, in short, feels miraculous.

By design, all their conversations reflected a significant difference—a crack at some layer of the earth. The students strongly and firmly disagreed about their main topic of discussion. But this crack didn't turn into an all-out rift. Thanks to receptiveness, even if they didn't persuade each other, their conversation didn't lead to those high-arousal, high-displeasure emotions in the upper-left quadrant that hold our hearts at knifepoint. It's a profound relief for my students to discover that disagreements don't have to be painful. They can be wonderful—validating, fun, productive, and rewarding—without causing harm.

The Receptiveness Mindset

The receptiveness recipe provides effective tools during disagreement, but it also works when any type of difficulty arises—when words, gestures, emotions, or motives collide. More than just a recipe, receptiveness is a *mindset* whose user remains keenly interested in understanding and validating their partner, even under stress. And with the right mentality, the ingredients of the receptiveness recipe should roll naturally off the tongue.

During difficult moments, we can have many goals at once: to learn about the other person's perspective, to validate their feelings, to persuade them to agree with our perspective, to maintain our conviction in our beliefs, to be open to their conviction, to avoid harm, to make a decision, to avoid making a decision, to run away, to keep our tears at bay, to look competent, to make sure the whole thing doesn't implode, and so on. Research has found that elevating some of these goals (especially aiming to *learn*) and squashing other goals (especially the desire to *persuade*) can help maintain a receptiveness mindset, which changes the course of the interaction. Those who focus on learning about the other person's perspective—who can let go of the instinctual impulse to persuade—are more likely to validate their partner's feelings, use receptive language, and, ironically, to ultimately persuade people to agree with them. In the words of my student, "We actually managed to persuade each other on a few aspects without even meaning to."

This kind of openness can be quite counterintuitive. When we have a strong, deeply felt disagreement with someone, our instinct is to try to persuade them that they're wrong and we're right—that what we want should happen instead of what they want. If it seems we can't win them over to our motives and beliefs, then we want to end the conversation as quickly as possible. Our desire to leave reflects a common tendency to think of beliefs as static and unchangeable.

Research by behavioral scientist Stav Atir and her fellow researchers suggests that people systematically underestimate how much they learn from everyday conversations, like what others care about and why. This is particularly true in the context of disagreement—we don't realize how much we could learn from a disagreeing mind. Instead we focus on being right and marshal the resources to prove it. We tend to assume that our partner is stuck in their ways—that they simply won't be open to having their beliefs altered by our different perspective. Recent research demonstrates that people persistently underestimate how much others are willing to learn about opposing views, and that when people disagree with us, we often dismiss them as "bad listeners" rather than as good listeners who hold different views.

Our tendency to underestimate other people's receptiveness may partly explain why many of us dread disagreement. If we don't believe anyone will learn anything from the conversation, it makes sense to avoid disagreements entirely; then when we inevitably do encounter disagreement, we think others aren't listening. Research has found that underestimating a counterpart's desire to learn strongly predicted that a conversation would result in conflict. When we think people don't want to learn, we treat them worse, and we lower our expectations about how well the conversation will go overall. So, it's a self-fulfilling prophecy: if you think people are out to persuade you (rather than learn from you), you will behave in ways that are more likely to lead to conflict, which inhibits actual learning and escalates discord.

Receptiveness runs contrary to these instincts, asking us to embrace a learning mindset rather than a persuasion mindset. When we are in a learning mindset, a disagreement doesn't have to be hostile or unpleasant. Instead, we focus on figuring out the sources of our differences, so we can resolve or accept them.

A learning mindset helps because it paves the way for a psychological process called *belief updating*. Suppose two people hold differing beliefs. Daniel believes that his seventy-five-year-old mother would be best served by moving to a nursing home, but his brother, Raymond, believes that their mother would be better off staying put in her house. It's possible for each brother to learn about the information underlying the other's belief, to judge the truth of it, and possibly to update his own belief based on new evidence. Only through this process of *receptive* disagreement can Daniel learn that Raymond worked hard to renovate their mother's house so she could live on one level. Raymond doesn't want that work to go to waste, and after paying for those renovations himself, he really can't afford to help pay for a move to a nursing home. Meanwhile, Raymond might learn that Daniel is incredibly grateful for Raymond's work on the house, believes the renovation will help with its resale value, and is happy to pay for their mother's move. As they revise their knowledge and beliefs, they inch closer to a different (more nuanced) understanding of the world and each other—and with luck, to agreement.

Adopting a receptive mindset, and expressing it through the receptiveness recipe, will encourage our partner to adopt a learning mindset, too. For those of you worried that receptiveness is equivalent to weakness, that it opens the door for others to walk all over you, fear not! Researchers have shown that receptiveness begets receptiveness. When you use receptive language, your partner is highly likely to reciprocate. Like so many aspects of conversation, partners show synchrony and accommodation—converging instinctively toward each other's level of civility in response. Though not impossible, it's quite difficult to be harsh to a chat partner who's being kind and receptive. This means we all have the power to start spirals of kindness and learning, even on the trickiest topics.

Identity

Let's go back to the brothers Daniel and Raymond, who are talking about their mom and the nursing home. During their conversation, it becomes clear that they have different levels of wealth, different values about DIY and manual labor, different levels of trust in eldercare facilities, and different beliefs about what it means to be a good son. Their conversations reach down to the deepest levels of the earth—to the differences in their identities. Though it can be roiling and high-pressure down there, most conversations plunge to the depths of our identities all the time, and in surprising and unforeseeable ways.

In 2013, I taught a negotiation case about a fictional quarterback named A. J. Washington who was negotiating his rookie contract with an NFL team. It was the second class I had ever taught as a professor—my first time teaching a course on negotiation. We discussed the nuances and complexities of Washington's contract, focusing on concepts like reservation value, outside options, the zone of possible agreement, contract contingencies, and so on. The discussion went great.

I had planned a big reveal for the end of class: the "A.J. Washington" case was based on superstar quarterback Tom Brady's salary negotiation with the New England Patriots. The colleague who wrote the case, Andy, had worked as the chief operating officer for the Patriots for many years, and that day he'd let me bring one of his Super Bowl rings to class. So when I made the big reveal, I flashed a photo of Tom Brady up on the screen in the classroom. "Enjoy the view, ladies," I joked with raised eyebrows and a cheeky grin. Then in a dramatic lunge, I pulled out the Super Bowl ring. The students gasped and lined up to take photos with it. I was crushing this new professor gig.

A few days later, I received an email from a group of three students who wanted to meet for office hours. I thought they would come to chat, like so many other students, about their job searches or interest in behavioral science or to seek my advice on some personal issue. They settled into the chairs in my office and started by telling me how much they were loving the class. Then they paused, and one of them said, "There is one thing we wanted to talk to you about." I was eager to hear. He went on to explain that someone had felt a bit sad about the Tom Brady moment. I was very surprised—confused, actually. They went on to explain that my comment, "Enjoy the view, ladies," had felt heteronormative, by which they meant that it addressed only the heterosexual students in the class. As representatives of the LGBTQ+ student group on campus, they wanted to bring it to my attention.

I was heartbroken. I viewed myself as a progressive twenty-eight-year-old and an ally of the LGBTQ+ community. I was shocked that I had made anyone feel alienated for their sexuality. I told them how truly sorry I was—I felt really bad. At the same time, a small (defensive) part of me also felt taken aback that such a brief quip—the word "ladies"—had inspired these three students to coordinate a private meeting to deliver this news. I had meant it ironically. Of course the students aren't all ladies, and they're not all romantically interested in Tom Brady. Was this offhand quip unironically offensive?

After some thought, I came to understand where they were coming from. What these students were asking for was recognition and validation—respect and acknowledgment—of their difference. My joke had made it sound like I assumed that all my students were heterosexual, that they would all find Tom Brady physically attractive. My offhand misrecognition of their identities had poked an invalidating barb into their core, one that LGBTQ+ students have to deal with all the time. And, even if this moment was fleeting, and the comment had been intended ironically, I knew the fracture could worsen if I didn't address it. They could feel ignored and dehumanized, and I too could come to question my own identity—as an open-minded ally and as a newly minted professor. This inflection point could lead to unpleasant, high-arousal emotions from the upper-left quadrant: frustration, anger, doubt, and contempt. Students might start to look for signs of my discriminatory bias everywhere, and I might start to look for their sensitivity and judgment everywhere—undermining our sense of trust and community. I'd seen this happen before in other courses—little moments between teachers and students, or among students, that seem disrespectful of others' nationality, gender identity, religion, native language, race, or any aspect of their identity. These fleeting moments can make a whole class break.

In my office, I apologized profusely and told the students how embarrassed I was to have made such a mistake. They told me that actually, most faculty—in fact, most people—made them feel this way, and I was the first professor they felt comfortable voicing it to. (*Thank goodness,* I thought, for my own sake, *but how awful for them.*) Returning to class the next week, I felt a sense of trepidation but also a sense of determination. I knew I could do better.

This story reflects a broader pattern: conversations across identity differences are risky. The less you know about your partner's experiences, the more you risk saying something offensive, and the more likely they are to distrust you or your motives. It all pushes the conversation toward the upper-left quadrant. My Tom Brady joke caused offense because I didn't have a full understanding of my students' perspectives and couldn't see that the joke I intended as ironic might be taken literally and truly affect them negatively. We risk making these mistakes all the time and in all kinds of ways—almost any topic or comment can feel linked to people's race, religion, sexuality, or heritage. Moments like these can create bigger problems even than disagreement because feeling that our core identity is misunderstood or disrespected is dehumanizing.

The Tom Brady joke-gone-awry isn't the only time I've poked barbs into others' identities—far from it. I fear it happens much more frequently than I even know, because barbs that pierce to a person's core can be invisible. With all these opportunities for things to go wrong, conversations across differences—from differences in understanding to differences in beliefs to differences in identity, which are inextricably linked at every layer of the earth—can seem daunting.

Perhaps it is no surprise that many people choose avoidance as a way out. Indeed, as recent studies by psychologists Jennifer Richeson and Nicole Shelton have demonstrated, larger numbers of people report that they are warier of conversation—feeling nervous, unsafe, and on the verge of being canceled—than ever before. Stunningly, the studies show that the desire to avoid conversation may be particularly strong among people whose identities are most threatened by conversations gone awry. For example, avoidance of conversations about race is especially prevalent among liberal whites who identify as antiracist. Living in a culture structured by a legacy of racism is, as the philosopher Tamar Szabó Gendler suggests, cognitively taxing for those who experience bias firsthand, as well as for those who wish to resist that legacy—for majority group members who desperately want to be allies of minority group members. Would-be allies can be the most avoidant because they're most worried about failing to adequately recognize someone's identity or about making things worse. It seems easier to opt out than to be part of the problem.

This avoidance isn't limited to politically charged identities like race. Going back to the brothers negotiating their mother's eldercare, Daniel and Raymond's identities were both wrapped up in being a *good son,* perhaps with slightly different ideas of what that means. Putting their mom in a nursing home when Daniel could take care of her in his home or Raymond could build a facility for her is not something a *good son* does. Grappling with these conflicted and painful parts of their identities might lead them to avoid each other, or to avoid this topic, or to have conflict on this topic, making things worse.

Getting Perspective

The question is how to lean into conversation across difference while keeping the conversation out of the upper-left quadrant. Like receptiveness to opposing viewpoints, talking across identity differences is a stress test of the Kindness maxim. Can we listen responsively, speak respectfully, and work to learn others' perspectives and identities *especially* when they're dramatically different from our own?

As children, we learned the Golden Rule: treat others as you'd like to be treated. This rule teaches us to be kind to others, even those who are very different from us. It's an appealing rule of thumb.

But can the Golden Rule help us overcome the difficulties of conversation across identity differences? In some ways, people want to be treated the same way you want to be treated. We all yearn to be validated, affirmed, heard, and understood, and to feel happy. But in other ways, they also want to be treated *differently* than you'd want to be treated, because everyone is different—they have traits, experiences, preferences, goals, feelings, and values that differ from your own. People like to feel respected, but respect might mean different things to different people.

Because of this, the Golden Rule can be misleading. When you're trying to be kind to others, you might be told, "Imagine it from their perspective" or "Put yourself in their shoes." It's a great idea in theory, but decades of research have documented the profound human inability to intuit the minds of others in practice. Put simply, people are terrible at putting themselves in others' shoes.

For example, psychologists Tal Eyal, Mary Steffel, and Nick Epley asked their study participants to imagine others' perspectives across a wide array of contexts. They asked them to predict another person's emotions based on their facial expressions and body postures. They asked them to judge fake versus genuine smiles in photographs, to predict when a person was lying or telling the truth in videos, to say what activity their spouse would want to do from a list of options, and to guess consumers' attitudes about different products. No matter what the researchers asked, they found no evidence that "trying to take the other person's perspective" helped at all.

The core issue is that, without direct access to others' minds, people tend to use intuitive strategies to guide their predictions. One common strategy is to consult the contents of your own mind, a practice psychologists call *egocentric projection*. Although our own perspective can sometimes be a good proxy for making social predictions, we rely much too heavily on easily accessible self-knowledge and fail to adjust for ways in which others' perspectives might differ from our own. Our egocentrism leads us astray: we overestimate the extent to which others share our own preferences (egocentric projection, false consensus); if we are experts, we forget how to communicate with novices (the curse of knowledge); we assume that others can see our inner feelings (the illusion of transparency); we judge people who are in a state of anger or anxiety, unless we feel the same way (hot-cold empathy gap); and so on.

But there is actually a fantastic way to know someone else's mind: *ask them*. In the last (twenty-fifth) study in their epic perspective-taking project, Eyal, Steffel, and Epley shared with the participants twenty different opinion statements from *Consumer Reports*—"I am an impulse buyer," "I am a homebody," "Television is my primary form of entertainment." They asked the paired participants to guess whether they thought their partner would agree or disagree with each statement. Guessing each other's preferences, feelings, and thoughts turned out to be woefully difficult (impossible). But when the pairs asked each other for their views directly, during conversation, they were able to genuinely understand each other's perspectives and report much more accurately on their opinions. It sounds obvious, but it leads to a profound insight. In the authors' words, "understanding the mind of another person is enabled by *getting* perspective, not *taking* perspective."

Conversation opens the opportunity to get others' perspectives—and it's especially important to take this opportunity when we're talking with *anyone,* not only those whose identity is different from our own. Even people who appear to be like us can have vastly different, less obvious aspects of their identity. Getting someone's perspective helps us know what they will perceive as disrespectful or offensive—it's the only sure way to avoid flubs like mine (and much worse). But it's also helpful to embrace perspective-getting as an end itself, as fodder for conversation. Meaning, we can see our lack of knowledge about someone not as a barrier but as an impetus for conversation. There's so much to discover about each other, and that's an amazing opportunity.

Other-Oriented

You've no doubt noticed that there are important similarities between perspective-getting and receptiveness. In the spirit of the Golden Rule, they both focus squarely on understanding and validating our partner (not on ourselves). They require an *other orientation.* They also both involve a learning mindset, learning the contents of our partner's mind, and learning from their experiences and knowledge, especially when they stand in stark contrast to our own.

Perspective-getting is helpful for those on the receiving end of bias, disrespect, or offensive comments, too. Instead of making hostile assumptions that offenders dislike, hate, or view us unfairly, we can ask about their motives. With great grace and love, my LGBTQ+ students came to meet with me privately—and in doing so, they could see that I didn't intend to exclude or shame them. Mine had been a fleeting comment, intended ironically by a loving person doing her best in a tricky and taxing conversational environment. (Groups are so hard!) Having their forgiveness was important to me for affirming my identity as an ally, which was crucial for our relationships, for the health of the classroom culture, and for the broader community at HBS. Their assertiveness and grace, combined with my receptiveness and resolve to do better, created a better—more inclusive, tighter-knit—community.

We can draw a how-to guide from the TALK maxims to aim for this type of perspective-getting through conversation:

- Ask caring questions.
- Show patience and grace if our partner doesn't really want to answer them.
- Make self-deprecating jokes about our own inexperience.
- Use respectful language.
- Validate their feelings.
- Listen generously and expressively.
- Show receptiveness to even starkly differing viewpoints.

While failures in perspective-taking have been shown to be the greatest barrier to conflict resolution overall, perspective-getting—through question-asking, validation, and receptiveness—provides the clearest pathway toward mutual understanding and connection.

The most important rule is not to be scared of conversations across differences. Even if you feel uncertain, having a conversation is usually better than avoiding it entirely. Research by psychologist Kiara Sanchez shows that conversations about race between Black and white friends can be incredibly rewarding, with friendships showing an increased sense of closeness even six months after the conversation. And a groundbreaking study by political economists David Broockman and Joshua Kalla in Florida found that ten-minute door-to-door conversations with transgender canvassers decreased transphobia and increased support for a nondiscrimination law.

Why? Because mere contact is humanizing. Conversation immerses people in the reality that every person is more than just one thing. Cisgender Florida residents were able to see, firsthand, that the transgender canvassers are funny, sweet, interested in art or basketball, make good eye contact, and ask good questions. They are the children of parents. The parents of children. And so on. For those who are willing to play the coordination game in good faith, conversation moves people from a monolithic example of one category (such as transgender) to a nuanced, complex individual who happens to also be transgender. That's the magic of conversation: we learn people's perspectives when we talk to them, no matter what we're talking about.

This is the ultimate goal of conversational kindness: to convey our recognition of people as they want to be seen—as complex, valuable individuals who are worthy of attention and care. Indeed, acceptance of difference is not an additional requirement for us to remember on top of our TALK skills—it is *the point*. Everyone we talk to is more than one or two or a few things, and every good conversation involves working mightily to suspend overly simplified stereotypical thinking and treat each human as highly nuanced and uniquely valuable.

Overheated

Sometimes, despite your best efforts—even if you've embraced receptiveness and perspective-getting with a learning mindset—conversation can get too hot. Your emotions start creeping into the upper-left quadrant, accompanied by automatic physiological responses like an elevated heart rate, stress hormones, or tears. These physiological symptoms are, at times, impossible to override.

What's more, emotions in that upper-left quadrant can emerge for a host of reasons that have nothing to do with your conversation partner. Maybe you have a headache or you're extremely tired. Maybe the last person you talked to made you really mad, and those feelings are coloring this next chat. Maybe you have a difficult presentation coming up later in the day, and you're feeling stressed. Wherever high-arousal, unpleasant emotions come from, receptiveness and perspective-getting won't always prevent things from heating up.

I once went camping with a friend at her family's woodsy cabin. One night, about twenty-five people—mostly her relatives—gathered around the bonfire. We were playing a game called Catchphrase, which requires players to give clues to their teammates so they can guess a certain word or phrase, like "Bohemian Rhapsody," without using any of the words in the phrase. ("Queen's most epic song—Freddie Mercury wrote it to sound like an opera!")

At one point, a player, Jill, got the word "baloney." She became visibly panicked, probably unsure what the word meant. She looked up to the sky, searching for help, and frantically landed on baloney's linguistic doppelgänger, shouting "Bologna," pronounced phonetically: *bo-LOG-na*. It was a funny choice made in a panicked moment. She even made an "I'm not sure I'm allowed to say that" face to acknowledge how silly it was.

Instead of letting it slide or correcting it and moving on, Jill's nephew John escalated it:

JOHN: You can't say *bologna* as a clue for *baloney*! That's ridiculous!
DORA (JOHN'S COUSIN): Don't talk to my mom like that!
JOHN: Well, it's freakin' ridiculous!
TED (JOHN'S DAD): Don't talk like that in front of your grandmother!
RACHEL (JOHN'S MOM): Well, he's right. It was a bad clue.
DORA: He can't talk to her like that!
JOHN: Are you mad at me for real?! For REAL?! She said *bo-LOG-na*! What the hell!

People stormed away from the bonfire. The five or six of us who remained were stunned, trying to make sense of what had transpired. My friend told me later that the baloney/bologna incident created rifts in the family that took years to heal, in part because it was a manifestation of other long-simmering family tensions. It wasn't *just* about the baloney. Still, a bad word choice made in a panicked moment spiraled into a blow-up. Even with good intentions, these types of spirals happen more often than we'd like.

Sometimes things get so heated that it's best to step away from the conversation completely, relying on the passage of time to cool things off and help everyone recover. But before things get that bad, there are other de-escalation strategies we can try.

There's a large literature in psychology focused on managing negative emotions writ large. This literature provides several concrete tactics we can use to manage our emotions when we feel things heating up during conversation.

The first tactic is to reframe negative emotions in a more positive light. When we feel anxious, angry, or afraid, we instinctively try to calm ourselves down. But making this kind of shift requires us to change our level of both pleasure *and* arousal—we have to move diagonally across the whole chart of emotions, from upper left to bottom right. My own research shows that it's more effective to reframe an unpleasant emotion as a pleasant emotion with the same level of arousal: to reframe anxiety as excitement, sadness as tranquility, or distress as passion. (The process is called *cognitive reappraisal*.) This simplifies the mental gymnastics: we're moving straight across the emotional chart, from left to right. With cognitive reappraisal, you're not trying to change the physiological symptoms (like slowing your racing heart, drying up your sweaty palms, or soothing your increased cortisol levels). You're just shifting your interpretation of those automatic symptoms—mentally reframing them as positive rather than negative.

My research shows that reframing negative emotions can have a powerful effect on your emotional experiences (you'll feel excited rather than anxious), on your behavior (you'll be a better conversation partner), and on how others perceive you (as more confident and competent). For example, in conversation, when someone asks, "How are you feeling about your upcoming presentation?" you might say "I'm excited" rather than "I'm anxious," which will help you focus on how things will go well rather than how they might go poorly.

During conversations, we can nudge our partners to reframe their negative emotions, too. Sometimes asking questions works—you're not asserting your reframes onto others but inquiring about their underlying feelings: "What is making you feel anxious? Could those things also be happening because you're excited? I can see you're thinking a lot about how things could go badly—what are some good things that might happen?" Therapists often help their clients with these sorts of reframes, but we can do it for each other, too.

In Chapter 4, we saw that the strategy of situation modification—shifting the context—can be used to raise the mood when it's sagging. But we can use that same strategy to de-escalate a mood. When my friend's new baby got upset, his mom advised him to "just add water." He was to take the baby and tiptoe outside into the rain, or plunk the baby into a warm bath, or walk along a beach or lake, or let them play with dishes in the sink, or spritz each other with silly spray bottles. Water, in almost any form, makes babies stop crying—the abrupt shift in context helps reset their mood.

For full-grown humans, the same principle of shifting the context works too. If we're not feeling good sitting down in a conference room, we might regather over lunch outside. If we start getting annoyed over someone's email, pick up the phone or meet face to face. When my married friends start to feel disconnected, they clap in sync and yell "Reset!" And context shifts needn't be particularly explicit. The options are endless. Just standing up, turning on the lights, lighting a candle, grabbing a drink, going for a walk, or inviting someone to join shifts the mood dramatically. Alas, we tend to forget that we have contextual choices once we've been sucked into the sustained engagement of conversation. But great talkers (and great stewards in groups), focused on the temperature of the conversation, are mindful of the immense effect even small environmental tweaks can have.

Sometimes simply acknowledging that a conversation is getting heated can be enough to halt escalation. Recent research by behavioral scientists Justin Berg and Julian Zlatev suggests that labeling others' emotions verbally ("You seem upset" or "It seems like we're feeling low") can foster interpersonal trust because acknowledging your partner's emotions shows that you're willing to give time and effort to meet their needs. My colleague Cynthia, an all-star teacher, will sometimes say in the middle of class, "I'm not loving the vibe in here. Should we hit the refresh button?" I admire this move—it comes off as perceptive, courageous, and caring, and it provides a great inflection point to shift topics.

Even more fundamentally, it can help us tremendously to know *why* we are feeling a particular emotion. The psychologists Yael Millgram, Matthew Nock, David Bailey, and Amit Goldenberg studied the extent to which people in their everyday lives understand the source of their emotions. Over a period of seven days, they measured how well the participants understood the source, gathering more than five thousand momentary assessments. They found that participants who had more knowledge of the source of their feelings used better emotional regulation strategies, like cognitive reappraisal and situation modification. They also found that these people were happier overall.

Pinpointing the source of our emotions can facilitate our ability to manage them and, from there, enhance our conversations. The same awareness can allow you to find out why your partner is feeling a certain way. With great topic management, question-asking, levity, and kindness, you and your partner can figure it out together. The TALK maxims help us plumb the layers of the earth and figure out our differences, while avoiding the dreaded upper-left quadrant.

Preventing Implosion

After my students survive the disagreement exercise—after they've seen how the ice cube of tense conversation can melt away with a talk partner who has also been trained in receptiveness—I ask them to go one step further: now do it for real. As an assignment, they must reach out to someone in their real life, outside class, with whom they tend to have conflict or are having a tricky disagreement. I challenge them to use their receptiveness skills to have a productive conversation and not to let things escalate.

In real-life disagreements, difficult conversational moments aren't always as obvious, explicit, or foreseeable as they are in our classroom exercise. In fact, many differences and tensions may be swirling beneath the surface, invisible, both to the conversation partners and to outsiders.

The conversations that the students record for this assignment range widely in topic—from reallocating household chores with a roommate to defining a relationship with a romantic partner, from addressing long-standing issues with parents to making strategic decisions with a business partner or negotiating a new job offer. Sometimes the difficult moments in the conversations they submit were foreseeable and anticipated, and at other times they arose suddenly, and the student scrambled to turn the naturally occurring situation into a homework assignment submission. The conversations are all fascinating, but one in particular sticks with me—a project submitted by a student named Emma.

To an outsider, the conversation Emma recorded with her sister, Samantha, sounds ordinary. Two slightly monotone voices trade thoughts on work and travel. The sisters themselves sound almost bored with their mundane chatter. It's what's underneath, under the crust of the earth, Emma explains, that made the conversation extremely tricky—and ultimately, extraordinary.

With no other siblings and only two years apart in age, Emma and Samantha were quite close growing up. After college, the girls moved into an apartment together to be near their father—he'd been in an accident, and they wanted to be nearby to help take care of him while he recovered. In the aftermath of his accident, they struggled. Worn down by daily worry about their dad, the sisters' once-close relationship became strained. Emma noticed that Samantha was exercising constantly and barely eating. When Emma confronted Samantha with her concerns, Samantha was defensive, rebuffing her offers to help.

Soon after this confrontation, Emma moved away for business school. For the last two years, they'd spoken only sporadically—to sort out logistical questions about their father (mostly through text messaging) and at family gatherings. Later, Samantha applied to business schools as well. On the occasions when they talked (mostly out of necessity), their conversations were stilted, filled with hurtful barbs, unresolved tensions, and—especially when Samantha was rejected from Emma's business school—new points of resentment. Emma explained that "most of what I know about her life now has been gleaned through my parents." The relationship was devoid of the joy they'd once shared. Emma worried they might lose touch forever.

Before calling her sister for a conversation for her homework assignment, Emma mapped out some topics she thought they might discuss: business school, New York City, Mother's Day, their parents' dog, advice about her career, and running. It was typical sisterly fodder, but land mines lay everywhere: Samantha had been rejected from Emma's business school. Talking about pets might remind Samantha of Emma's dog, Daffodil, which Samantha believed she should get rid of. They'd failed to coordinate Mother's Day gifts together for a couple years. And talking about exercising might dredge up major sensitivity around Samantha's health issues. Emma found it hard to think of any topic that was guaranteed to be safe—one that wouldn't touch on painful shared experiences or sensitive parts of their identities. Based on her learning in class, though, Emma hoped that this time she could make any topic feel safe.

"Thank you for agreeing to do this," Emma starts in the recording.

"No problem. I'm just—driving." After initially saying no, Samantha has agreed to talk on the phone during a two-hour solo drive between cities.

They chitchat about the traffic and the weather, idling temporarily at the base of the topic pyramid, then veer into one of Emma's prepped topics: business school. They stay on this topic for some time. Emma is nearing the end of her MBA program, and Samantha will start at a different business school in the fall. Though it's an obvious point of common ground, it's also a charged topic—a painful point of difference—as Samantha's rejection from Emma's school is still raw.

Rather than talking too much about her own experiences, Emma asks Samantha what she's looking forward to. Samantha says she's excited to start a new chapter and hopes to travel a bit with new business school friends. Despite Emma's focus on Samantha's perspective, at the mention of travel, she can't help but share that she's leaving for an amazing destination—Mount Kilimanjaro—in a few days.

"Alone?" Samantha asks with surprise.

"No. We have like an outdoors club that I'm a part of. . . . They had set up this trip to Mount Kilimanjaro."

"You're part of an *outdoors club*?!" Samantha asks, dripping with disbelief.

Emma is wounded by Samantha's comment. She understands the skepticism—she surprised herself by joining the outdoors club two years ago, and the same question from almost anyone else wouldn't have felt hurtful. But coming from Samantha, the comment highlights their shaky ground. It makes Emma feel misunderstood and, well, abandoned. She doesn't feel seen or known—quite the opposite. Due to their distance (geographical and emotional), Emma feels like Samantha doesn't quite understand who she's become over the last two years. Instead of being excited and amazed by her big trip, her sister's lack of validation has jabbed a barb deep into Emma's identity.

This is a difficult moment. Emma's knee-jerk reaction is to defend herself—to explain how she's changed since she moved to school. How she's a new person whom her sister doesn't really know. To accuse Samantha of losing touch.

But Emma remembers the receptiveness recipe. Instead of lashing out and getting stuck in a defend-and-attack conversational loop, Emma responds, "I know! I still don't like to camp, but I do like to hike."

She goes on to share that, although she was skeptical about it at first, too, the group has been "amazing," taking her to Egypt, with a cruise down the Nile. She's confident Samantha will find similar joy in unexpected places, too. "I'm sure they'll have these at your school, too. There are a few organizations . . . not everyone but probably like half the students join certain clubs."

"Egypt's not exactly on my bucket list, but I'm kind of open to whatever." Samantha's skepticism is still there, but perhaps dissipating.

"Yeah, Egypt and Kilimanjaro weren't on my bucket list either," Emma says, affirming Samantha's skepticism. "But Egypt was amazing, and I'm looking forward to Kilimanjaro now, too." She then turns the focus back to Samantha: "Do you have anywhere on your travel list for the next two years?"

"I do want to go to Thailand."

"Oh yeah, I've heard Thailand is awesome," Emma offers with positivity.

"Yeah. Honestly, I'm grateful that I won't, like, need to take out loans to do these things," Samantha admits. "I'm so lucky to have a company that's sponsoring my degree."

The topic ends in a hopeful, grateful place. In reflecting about this part of the conversation, Emma writes that she usually worries that if she doesn't defend herself, then she will feel or seem weak—like she hasn't stood up for herself. But in this moment of atypical receptiveness, she felt the opposite: "I felt more secure than ever about my interest in the outdoors club—I think precisely because I didn't feel the need to aggressively defend it." Emma describes this moment, despite its benign appearance, as extremely important. "I made a conscious effort to avoid or diffuse potential anger—my own and Samantha's—even when Samantha made a comment that would have typically made me very defensive. This wasn't easy. But I feel really proud that I did it."

Emma notes that after navigating this tricky moment, she "never made a conscious effort to switch topics. . . . But I did make a conscious effort to ask a number of follow-up questions . . . which did a great job of increasing enjoyment. If you ask a lot of follow-up questions, you don't need many topic switches." Yes, Emma, yes!

In the recording, the sisters go on to consider Emma's job offers and where life might take her next. "You have to be happy with yourself first, then good things will come to you," Samantha says.

Emma's taken aback Samantha's maturity. "When did you learn all this?" she cajoles warmly.

In her reflection about the conversation, Emma realizes that Samantha used several tools from class too, perhaps unconsciously. When she disagreed with Emma, she hedged her claims and expressed her listening verbally. Emma noticed and appreciated Samantha's kind moves more than ever.

Despite the land mines they'd faced, they reached the end of the call on "shockingly" good terms. Emma feels relieved and encouraged and expresses this feeling to Samantha at the end of the call. "Let's talk again. It's been so nice to reconnect," she says. It's the kind of thing someone says when they're trying to mark the significance of an event.

To Emma, this conversation is incredibly important—not just as the fulfillment of a class assignment but also as a momentous olive branch offered to her sister.

Samantha teases her back, "Well that's awfully formal." They both laugh. After forty minutes of back-and-forth, it's the most notable moment of shared laughter, filled with mutual feelings of relief.

. . .

We tend to think of conversations as having a certain drama to them, but often that drama is hidden—held inside our hearts and minds, unspoken. Transcripts of conversations often don't reflect the emotional tenor of the spoken exchange because in many cases the drama is in our heads—in our reactions to what's said or our fear that our response has been wrong or inadequate. Emma and Samantha's phone call is a great example. To an outsider, their call sounds relatively unremarkable. There are no overt disagreements. No noticeable differences between the sisters' language or emotions. Their voices are similar, and even their tendency to pause between conversational turns is well matched. There are a few moments of repair when the phone connection wobbles, though perhaps fewer than most other people might experience during a standard call while driving. It's the kind of conversation we've all overheard on a train or at an airport.

And yet at every turn, both women had to make critical decisions about how to respond, informed by their shared past and personal histories. Though ordinary on the surface, Emma's reflection about the call reveals its remarkable depth: "This was the most rewarding project I've completed in business school. Our relationship was broken. While it's not entirely fixed, this gives me hope that we are moving in the right direction."

Toward the end of the call, Emma invites Samantha to stay with her during her graduation, just a few weeks away, after the trip to Kilimanjaro. It's an invitation Samantha is surprised to receive, but a few days later she accepts. Emma is thrilled and optimistic: "I hope this is the beginning of a much more rewarding relationship."

THREE KEY TAKEAWAYS FROM CHAPTER 7

Difficult Moments

- Differences—in words, emotions, motives, and identities—can all cause **moments of difficulty** in conversation.
- Use the **receptiveness recipe**—acknowledge, affirm, validate, hedge, aim to learn—to engage with opposing viewpoints.
- Use **situation modification** and **reframing** when emotions get hot.

CHAPTER 8

Apologies

TASHIRA AND DRU SIT A few feet apart on a velvet sofa, looking up to the ceiling, sighing. Across from them, Dr. Orna Guralnik, her dark hair pulled neatly into a ponytail, fixes her gaze intently on her patients. The room is wrapped in caramel-colored grasscloth wallpaper. It's a soothing place, but what happens within those four walls often isn't. With cameras and microphones hidden among rows of books and objets d'art, their therapy sessions are being filmed for Showtime's documentary-style television series *Couples Therapy*.

Tashira was already a single mother when, after a couple months of dating, she became pregnant with Dru's child. Now they're raising two young children together, trying to make it work. It feels like a roller-coaster ride, Dru said in their first session—fast and bumpy. Tashira wishes they could slow down, but everything feels out of their control. She sleeps with her two little boys in one bedroom of their New York City apartment, while Dru sleeps alone in another room. Tashira is fine with this setup because she's working hard and needs to retreat to her room. But Dru doesn't like it—he feels rejected and alone. Their disagreement about sleeping arrangements isn't their only problem. It's one of many ways Tashira and Dru aren't seeing eye to eye.

Dru opens this, their second therapy session, with an accusation. "She's not a good communicator," he says, "especially when something is bothering her."

Tashira leans forward, not yet disagreeing, but skeptical about where Dru might be heading.

Guralnik asks Dru to give an example.

"Okay, so this morning, there was some stuff on the table. Paperwork or something. It's been there for a couple of days. Now it's driving me nuts." He'd left the papers in an obvious spot on the table for days, hoping Tashira would take care of them, but she hadn't. "Instead of me saying anything negative, I bunched it up and smiled. Like 'Look what I put here for you.'"

Instead of complying or thanking him for his nonbarky (passive-aggressive) reminder, in that moment, standing across the table, Tashira clapped back. Dru recalls, "She threw something in my face like, 'When I do the laundry folding, you don't put it away right away.'" This response, Dru says, made him upset. "Why do you have to sit there and fight me? Why does it get to a point of now we have to face off?"

Guralnik redirects to Tashira, who picks up the story.

"I grabbed the stuff and said, 'Okay, I'm sorry, I won't leave my things on the table anymore.'" Then she went into the bathroom to finish getting ready for work. At that point, she says, the issue was resolved. "It's over. It's done."

But it didn't feel resolved to Dru. "The way she apologizes is basically a silent 'Shut the fuck up. Leave me alone. Get out of my face.'"

It's easy to see how Dru felt like Tashira hadn't apologized at all—how her pseudo-apology may have made things worse. Tashira explains her underlying frustration: "When it comes to house things, I have asked you a million times to do a lot of things, and you don't do them. And I just gave up asking you."

Keeping a house—and a life—in order is relentless work, especially for those who have two jobs and babies. It's clear to Tashira that Dru hasn't apologized for not doing his fair share around the house, either. She didn't really want him to move in, and now he's there, not carrying his weight, guilt-tripping her about some papers. Meanwhile, he's made her think she's not allowed to express how she feels because he thinks she's overly negative. So instead, she chews on her resentment in silence, ready to clap back whenever she's pushed too far.

When we dissect the many conversations that make up a relationship over time, we find that what matters is not really the pile of papers on the table, or the laundry someone failed to put away that day, or any *one moment*. Rather, each little moment is an example of something bigger—examples of what the partners need from each other but feel they are not receiving. Almost nothing we have to apologize for is about the thing itself. It's about whether we're failing to fulfill each other's needs.

If kindness means giving our partner what they need, then failing to give it to them, especially when they believe their needs are obvious, causes conflict and harm. And like papers stacking up on a table, our micro-harms accumulate over time.

Acts of Trust

Every relationship you've had with every person you've ever known has a shape that formed over time—a trajectory. The *New Yorker* cartoonist, writer, and columnist Olivia de Recat captures the idea of relationship trajectories over time with what she calls "closeness lines." Imagine two strings that start at "hello" and stretch to the end of your lifetime. There is a unique string for every person you encounter in the world.

Comments from de Recat's many readers attest to the stirring simplicity of her illustrations. "The parent one brought me to tears," wrote one reader. "The subtle curves of first love are telling a story," wrote another. I too find her closeness lines poignant and thought-provoking.

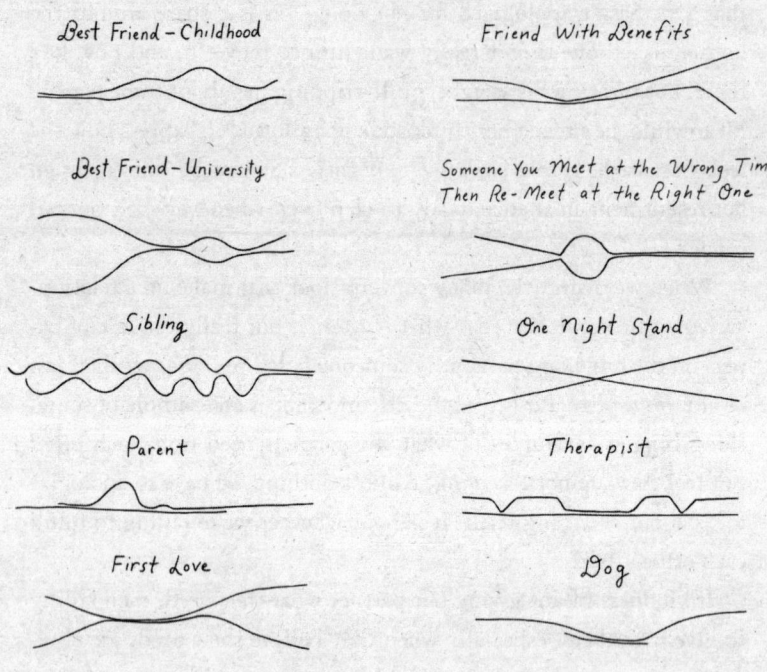

Olivia de Recat, Closeness Lines, 2019

But I am more interested in what you *don't* see on De Recat's closeness lines. Where, on those silent stretches of string, are the *knots*? The points in time and space where we converge for a moment of talk with a sibling, first love, parent, one-night stand, best friend from childhood, best friend from college, and so on? Our conversations are like itty-bitty knots spread out along the way, whether randomly or at regular intervals, close together or far apart. We may or may not even remember which conversations kept our closeness lines in touch, which ones pulled them closer, and which ones sent them hurtling asunder.

Tashira and Dru's closeness lines tell their story, too. They were flung together by an unexpected pregnancy—so fast that they could barely hang on. Though they're living together and raising children, their lines are wobbly, an uncomfortable holding pattern at some distance apart, threatening to diverge.

Our closeness lines wobble and drift apart for many reasons. We move on, or we move away. Life pulls us in different directions. Perhaps the closeness was contingent on a particular moment in time. Our lives have different seasons, and the people in them change along with them. Sometimes, though, a disagreement or difference can send us hurtling away from each other—causing an abrupt and unwelcome weather event in our lives. Following a rift, the passage of time can bring closeness lines back together, especially for people who are bound by obligatory exposure to each other, like some work colleagues or family members. But the passage of time isn't always enough—or optimal. Sometimes conversation is required to diagnose what went wrong and to try to fix it.

When we as partners have done harm—and receptiveness, responsive listening, and emotion regulation haven't helped enough—apologies are one of the most powerful tools we have in our conversational toolkit. At these moments, not just a conversation but a whole relationship, like Tashira and Dru's, may be under threat. Apologies—and the omission of them—are pivotal inflection points along our closeness lines, which is why good apologies aren't just good in the short term. They can strengthen relationships, creating a virtuous cycle defined by a supportive equilibrium over time. Meanwhile, bad apologies can degrade our relationships, creating a vicious cycle of resentment. Bad apologies, like Tashira's snippy, sarcastic "Okay, I'm sorry," are themselves harmful violations. But so too is the absence of an apology, like Dru's.

When I spoke to their therapist, Dr. Guralnik, she explained, "Apologies can be incredibly powerful, because they are conversational shorthand for showing two underlying things: (1) that someone understands the world through their partner's eyes, and (2) that they are willing to take responsibility for harm. The absence of those things—understanding and taking responsibility—really pisses people off. It's what makes us feel desperate to receive apologies. We're all desperate to be understood and for others to take responsibility."

At their core, apologies are about restoring trust after it's been broken. Luckily, the TALK maxims have trust at their core. They guide us to choose topics, ask questions, lift the mood, and show kindness, so that our partners might come to trust that we care about them, understand them, and want to help them achieve their goals. The maxims help us, even in a difficult circumstance like the kerfuffle in a group or a thorny conflict, communicate successfully and build trust. In fact, the whole right side of the conversational compass—highly relational pursuits like having fun, exchanging advice, being honest, soothing each other, avoiding awkwardness, and reminiscing—are goals that build, reflect, and help sustain trust over time.

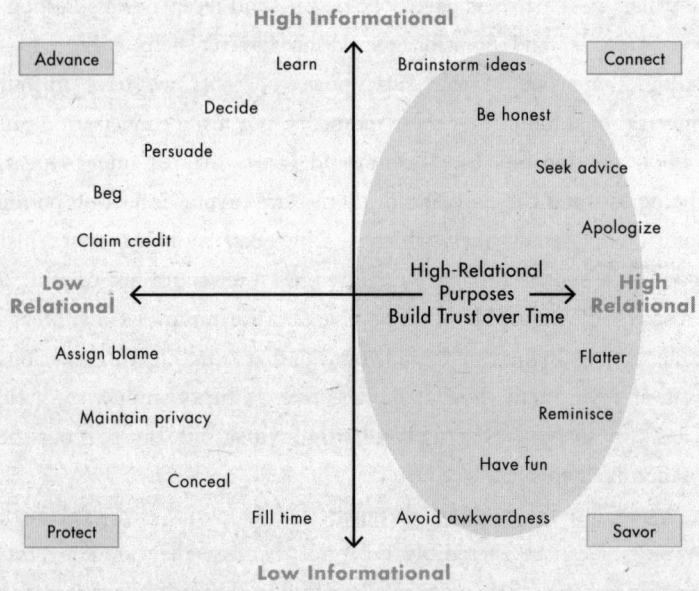

Establishing trust (and its doppelgänger, psychological safety, in groups) is a key aspiration of the Kindness maxim in particular. We show kindness to our partners by using respectful language and listening attentively, not only for the sake of kindness but also to build trust, so they see us as the type of person who will be kind to them in the future, too.

Apologies are an apex conversational skill that combines and goes beyond everything we've learned in the book so far. Apologizing well is tricky because we usually need to do it during moments of intense relationship stress—during times of difficulty in our lives, in the throes of pain, or after a transgression, when something feels broken. And in aiming to heal relational strain, apologizing requires us to deeply examine ourselves, our behavior, and our motives. It requires us to be our best selves.

To Apologize or Not

It's hard to say you're sorry. Despite the power of good apologies, most of us are reluctant to make them because there are risks. An apology hopes for forgiveness, but our partner doesn't have to forgive. Apologizing makes us feel exposed. Plus, despite obvious harm, we still may not want to concede the point. We get stuck focusing on our own perspective, worried about maintaining a positive view of ourselves and wanting to be *right,* rather than working to understand our partner's perspective, which requires perspective-getting and humility (a relentless willingness to be *wrong*). While we might understand a transgression—and our role in it—a certain way, our partner is likely to see it differently.

Children, all the time, give us examples of an excessive focus on our own perspective. They are often wrong, they want very badly to be right, and they are relatively incapable of considering others' perspectives. When my oldest child, Kevin, was a toddler, he was filled with intense energy and big ideas that he couldn't express with his words, and he was *so frustrated* about it. For about a year, his intense feelings made him aggressive. It wasn't the idyllic picture of motherhood I had imagined before having children. Often enough, I got overheated, too—reciprocally angry, frustrated, filled with rage, ready to yell. And trust me, sometimes I did yell at Kevin, even though I knew it wouldn't help. It only escalated our conflicts and made me feel guilty and embarrassed.

Once, I was picking him up from the ground midtantrum. He arched his body and flung his head backward, right into my face, breaking my nose. My eyes welled with tears—from physical pain, but also from frustration. I yelled "Kevin!" (picture the mom in the movie *Home Alone*), plopped him back down, and walked away to assess the damage in a mirror. I was hurt physically but also emotionally. I was upset that I'd created a person who was capable of hurting someone—that I hadn't yet taught him not to. I suspect he understood and felt bad that he'd hurt me, but he didn't know how to express it—and perhaps he cared more about getting his way than my pain.

To a less obvious degree, adults are the same way. We overfocus on our own perspective and goals. Even when things have gotten out of hand and our partner is hurting, we may still be unwilling to concede—unwilling to prioritize their needs over our need to be *right*.

When we think about the apology process, it's easy to decide that it's just not worth the trouble. Some might even view apologies, explanations, and promises to change as meaningless "cheap talk," with no real value compared to more substantive, tangible reparations—money, concessions, favors, acts of service—to restore trust. Even Dr. Guralnik admitted that her feelings about apologies are "mixed" because apologies can seem "redundant" if two people are able to show understanding for each other and take responsibility for their transgressions without apologizing.

But on this point, the science is clear: in conversation, apologizing is always better than not apologizing. Research shows that those who apologize are ultimately viewed as having *higher* status than those who don't. Blameworthiness doesn't matter as much as we think it might, and verbal expressions of apology are incredibly consequential because they acknowledge harm and show that the apologizer believes the relationship is *worth fixing*. That their partner is *worthy of care*. Though reparations can also help, apologies alone are extremely consequential. In fact, there is not a single study of conversations that has found not apologizing to be better than apologizing.

Perhaps you're thinking, *Isn't apologizing tantamount to admitting I was wrong? What if I wasn't wrong? What if my partner was being ridiculous, mean, offensive, or unethical, and they're wrong?* Good questions. Even when we view someone else's beliefs or motives as ridiculous or unacceptable, to them, they are not. They have a different perspective, a different reality. If you care about someone, then you need to help them pursue (or revise) their motives, no matter how ridiculous they may seem. They may change what they believe or care about slowly over time, but probably not within one conversation, and almost never within one heated conversation.

Or you may be thinking, *Yes, but couldn't saying I'm sorry be admitting my guilt? Couldn't I be sued or convicted if I admit to causing harm? Am I giving something up here that I shouldn't? Don't I look worse if I admit I was wrong?* All good questions, and some data can help us answer them. For years, when medical professionals made mistakes that hurt or even killed patients, they were advised not to apologize because it might make them or their hospital liable for a malpractice lawsuit. But then some hospitals began allowing doctors to offer apologies to patients and families, and some even made apologizing mandatory. Research has shown that these policies *reduced* the likelihood of litigation. People are more likely to sue others not for being wrong but for being cruel.

Instead of thinking of apologies as embarrassing admissions of incompetence, negligence, or guilt, or as unpleasant penances that should be avoided unless absolutely necessary, we must think of them as the most powerful opportunities to show love and strengthen trust.

The Power of Apologies

Apologies are powerful, so powerful that you don't even have to have done something wrong for them to have an impact.

In November 2010, Maurice Schweitzer, Hengchen Dai, and I sent a research assistant, Tim, to a large train station in Philadelphia to help us study trust between strangers. We asked Tim to approach people and ask if he could borrow their cell phone. The request was risky for the commuters, as Tim could easily have absconded with their device. Here's the trick: we sent him out only on *rainy* days, which enabled him to apologize—"I'm so sorry about the rain!"—or not, before making his cell phone request.

When he made the request without the apology, 9 percent of the busy commuters handed him their cell phone. But when he said "I'm sorry about the rain" before making his request, the percentage of people who handed it over jumped to 47 percent. That's a fivefold difference. And he was apologizing for something (the rain) that was clearly not his fault.

We then measured trust behavior in economic trust games using other types of superfluous apologies—for a partner's delayed flight, for heavy traffic, or for bad luck (all things that were clearly outside of the apologizer's control). In every case, apologizing increased feelings of trust and trusting behavior—those who apologized were rated as more trustworthy and were better liked, and their partners were more likely to choose them for the next round of the game. The power of conversational apologies comes from recognizing a partner's struggle, however small, and from showing that you care—even when it's obviously not your fault.

Relationships, then, that develop healthy apology habits—pairs who apologize easily and receive apologies with gratitude and forgiveness—are likely to be stronger (more accepting, trusting, and sturdy) than those who don't. Research in 2012 by psychologist Karina Schumann tracked the apology behavior of sixty married couples.

She asked them to report their relationship satisfaction and complete daily diaries in which they reported the transgressions made by their partners, the apologies made by their partners (including their perceptions of sincerity), and their willingness to forgive their partners. People who were highly satisfied with their relationships and who apologized to their partner were more likely to receive forgiveness. Those in sturdy relationships were more forgiving after their partner's apologies, because they viewed them as sincere expressions of remorse.

In a different study, Schumann and her fellow researchers found that some partners have a higher tendency to apologize than others, and that apologizing frequently didn't devalue the impact of their apologies. The only negative outcome Schumann observed was when apologies were seen as low quality—insufficient, halfhearted, insincere, off-base, or self-interested. Taken together, these results suggest that the key is to apologize frequently—and to do it well.

The Worst Apology in History

On April 20, 2010, forty-one miles off the coast of Louisiana, an oil rig over an underwater valley on the continental shelf known as the Mississippi Canyon experienced a catastrophic blowout. A surge of natural gas blasted through a recently installed concrete seal, coursing up the rig's riser to the platform above the water, igniting an explosion that could be seen from forty miles away. The explosion killed eleven workers and injured seventeen more. Over the next two days, the severely damaged rig, the Deepwater Horizon, burned uncontrollably and then sank into the ocean.

BP, which owned the damaged well, estimated the volume of oil that escaped to be about 1,000 barrels per day, but U.S. government officials thought it peaked at more than 60,000 barrels per day. The leaked petroleum formed an oil slick extending across more than 57,000 square miles of the Gulf of Mexico. It crippled fishing and tourism industries all around the gulf, leaving an estimated twelve thousand people unemployed, and it wreaked incalculable damage on the environment and wildlife. It was the largest marine oil spill in history.

BP chief executive Tony Hayward emerged as the public, and increasingly problematic, face of the tragedy. He initially downplayed the spill, stating on May 13 that its environmental impact would likely be "very very modest" and calling it "relatively tiny" compared to the size of the ocean. By May 28, he had changed his tune, calling the spill an "environmental crisis and catastrophe" in an interview with CNN. He was alternately dismissive and circular in his responses during media interviews. Perhaps his lowest point came on May 30, when he offered an apology for the spill. "We're sorry for the massive disruption it's caused to people's lives," Hayward said. "There's no one who wants this thing over more than I do. I'd like my life back."

And with that, Hayward earned a place in the annals of apologies. It is surely one of the worst apologies in history. He downplayed the harm; the loss of life, and the destruction of more than twelve thousand jobs in fishing and tourism, is more than a mere "disruption." And he missed a crucial opportunity to express remorse and culpability for the devastation caused by BP under his leadership. The apology was made worse when details came out about his multimillion-dollar paycheck, as did the fact that he'd taken time off during the cleanup to watch his co-owned yacht, *Bob*, race around the Isle of Wight. His botched apology required subsequent apologies posted on Facebook.

Why? Because Hayward had committed the cardinal sin of apologies: he made it about *himself*.* Decades of research show that effective apologies focus on the experience of the recipient, not the apologizer. Even a hint that the apologizer is using the apology to rectify their

---

* In October 2010, after enough time had passed for him to to process the events, Hayward gave a better apology: "The Gulf of Mexico explosion was a terrible tragedy for which—as the man in charge of BP when it happened—I will always feel a deep responsibility, regardless of where blame is ultimately found to lie."

own feelings, to unburden their guilty conscience, to wash their hands of a tricky situation, to express their frustration, or to try to move on—rather than to tend to the recipient's feelings—can undermine the effectiveness of the apology.

To apologize well, we can't make it about ourselves. Instead, we should focus on our partner's feelings in the moment—and stay on the topic at hand—without extrapolating to other issues or clapping back with our own frustrations. An apology should be an expression of empathy—a way to say "I realize something bad happened to you, or something has upset you, and I'm sorry." Like Tony Hayward, Tashira made Dru's plea about the clutter on the table about herself. She used it as an opportunity to air her grievances about his lack of housework—"When it comes to house things, I have asked you a million times to do a lot of things and you don't do them"—rather than focusing squarely on his frustration in that moment. (My husband calls this nasty habit "snowballing," and like everyone, I struggle with it, too.)

The first step is to seek to understand the other party's experience. In Hayward's case, he hadn't yet recognized the full extent of the damage—and BP's role in it. If he had understood and felt the magnitude of it, his own discomfort would likely not have felt so important. Research by Roy Lewicki shows that when you acknowledge responsibility, you show that you understand what your partner is going through—that you understand the extent of the harm you caused.

Sometimes that extent is not immediately obvious until we ask our partner to explain their perspective—until we sincerely seek to understand and engage in some sense-making back-and-forth. We must sift through the layers of the earth to identify where our differences lie—where the hurt happened. During that process, it's important to continually remind our partner that we want to understand their perspective so that we can figure out a good path forward.

Lauded drummer and frontman of the hip-hop band The Roots, Ahmir Khalib Thompson, aka Questlove, gives us a good example of this perspective-getting process. He narrates how he misunderstood but has come to better understand how he made some of his fans feel. Here, in an Instagram post, he responds to his Japanese followers who helped him understand how a joke he'd made had hurt them:

> Unfortunately, I've offended my Asian brothers and sisters with an IG post which I made during my recent tour of japan. In that post, I likened a japanese department store employee's vocal intonation to that of a (church) deacon speaking in tongues. Clearly, I didn't intend to offend anyone (asian or otherwise). Clearly, I "thought" that comparison was funny-cute . . . and clearly I thought wrong.
>
> In hindsight it's easy to see how my post was yet another example of the ugly american flipping yet another ugly/racially/culturally insensitive script. So, let me make this abundantly clear . . .
>
> THE [STUFF] THAT I SAID WAS DUMB (PERIOD).
>
> —look. I'm a human being and dumber yet, I'm a public figure. If you're lucky enough to be either of the aforementioned, then not only should one stay clear of saying or writing hurtful things, one should actively work against feeling comfortable, thinking hurtful thoughts. Given that black culture consistently finds itself at the butt end of so many offensive "outsider" jokes, I should be way, way more sensitive (after all, who's zooming who). I for one, should never allow my cultural bias to take precedence over my "examined life" (clunker be damned). I know the whole kinder and gentler thing reeks of a self-serving political correctness, but eff it, it's "all me."
>
> So, here I am once again, publicly coming to terms with some more of my stupid "say, say, say" ish, allow me to ask for forgiveness and understanding from anyone that I've offended. I will be better in 2014 (I promise).

Questlove's apology is pretty darn good. He expresses care for his fans, acknowledges his wrongdoing, and walks us through his realization process, which seems humble in its humanity. He self-castigates, shows cultural awareness about the oddness of public apologies, and promises to do better in the future.

Still, he too has a stumble near the end: he makes the apology about himself when he asks for forgiveness. Requesting forgiveness is a tempting component to include in an apology—when we've done something wrong, we are hungry for forgiveness. But even if seeking forgiveness is a core goal, it's a self-serving one. And asking your partner for forgiveness in the same breath as you're making an apology makes it too much about yourself—you're asking your partner to make *you* feel better, rather than focusing on making *them* feel better. It may be best to avoid asking for forgiveness. Let your partner decide on their own time if and when they're ready to forgive.

I've Changed

In the United States, after serving a minimum sentence for a crime such as battery, theft, rape, or murder, people who have been incarcerated may become eligible for parole. If granted, parole can cut the time they spend in prison by more than half. During the parole process, a group of officials—the parole board—tries to figure out if the incarcerated individual is ready to be released, balancing many factors, like the seriousness of their crime, the amount of time they've spent in prison, their behavior in prison, and so on. Part of this process is a parole hearing—a conversation between the incarcerated individual and the parole board.

Though parole hearings are much more intense than most of our everyday encounters, they provide something incredibly valuable that our everyday interactions often don't: a clear-cut outcome. For each hearing, we know whether the incarcerated person was released from prison afterward. This means we can link the elements of their language during the parole hearing, including any apologizing they may have done, to the parole hearing's outcome. Together with psychologists Grant Donnelly and Hanne Collins, I've transcribed and analyzed over three thousand parole hearings conducted in Nevada and Kentucky in 2017—the largest sample of parole hearings ever compiled and transcribed, and the first large-scale investigation of apologetic language in natural dialogue.

Decades (maybe centuries) of research on apologies has theorized about the different elements of apologetic language, and the new science of conversation is helping us figure out which theorized elements people tend to use—and should use—in real conversations. Roy Lewicki's research points to seven linguistic features: statement of apology ("I'm sorry"), expression of remorse ("I feel so bad"), offer of restitution ("I owe you"), self-castigation ("I'm an idiot"), request for forgiveness ("Please forgive me"), promise regarding future behavior ("I will be better next time"), and explanation for the transgression ("I did it because I was angry"). I agree this is a useful breakdown to think through the different aspects of apologies, and we saw several of them in Questlove's apology.

We asked an army of research assistants to annotate the parole conversation transcripts, indicating which turns of conversation included which of Lewicki's elements of apology. We found, as you'd expect, that when people up for parole give explanations for their crimes like "I was at the wrong place at the wrong time and just got railroaded," they are much less likely to get out of jail. So too with "I was out of my mind, wasn't thinking properly, and was into drugs. My wife was dying of cancer and I got out of control." In Nevada, giving an explanation of the crime had roughly the same *negative* impact on getting a prisoner out of jail as their having committed a prior felony. Reminding the board about the circumstances of your crime was as bad as actually having committed another one.

But one of the apology elements, we found, had a powerful effect on likelihood of release, one as powerful as being a woman (compared to a man). It was the only element that had a strong positive link to prisoners' getting paroled: *promise to change*. Just reading these promises, you can feel their power:

> "I intend to never have another offense. I'm 65 years old, I've matured, I've changed, I am a different person."

> "I haven't used drugs since I have been here. I have been clean and sober, and I have been working and maintaining a good program. That is all I have been trying to do. I have a grandson now. I have things that are very important to me that I need to get out.... It's a wrap. I understand that. I just want to clean up what I owe, get back into society, start working hard, and I know that drugs are a dead end."

> "It's just not an option for me to reoffend again, I won't allow it. I won't make any choices that are anything other than what I know I need to do to stay out and victim-free and live a healthy life."

While we can include many elements in our apologies, promising to change is in a class of its own. The inmates' promises to change often focused on their concrete plans once they get out, like attending weekly AA meetings, going to church, or living with upstanding, loving family members.

Promises to change pull the recipient's mind to the future—they signal who we can be rather than who we were or how our trust was ruptured. They soothe the recipient, who is looking for assurances that reduce the distress and uncertainty caused by the violation. And because promises (like all language) are "cheap talk," the more concrete and realistic these promises are, the better, including plans for reconciliation and how we'll avoid making the same mistake again.

Or better yet, we can try to give behavioral evidence that we've already started to enact positive change.

Promises to change aren't effective just in parole hearings. Behavioral scientists have studied the trajectory of trust breakdown and recovery by asking participants to play a "trust game" that mirrors the trust breakdown and recovery processes we encounter in our everyday lives. In one study, the trust game required each player to pass money back and forth with another (simulated) player across several repeated rounds. At each step, the player with the money in hand decided whether to take it for themselves or hand it back over to their partner (which would dramatically increase the size of the pot). Passing money to your partner—like handing over a cell phone to a stranger in a train station—is a highly trusting behavior because you don't know how they'll respond.

In all the pairs, the simulated partner violated the participant's trust by keeping the money for himself in the first two rounds of the game. Trust was broken. Then the simulated partner tried to restore trust in several ways: by making an apology without a promise to change ("I really screwed up, I shouldn't have done that. I'm very sorry I tried taking so much these last two rounds."); by making only a promise to change without an apology ("I give you my word. I will always return the money every round, including the last one."); by making an apology plus a promise to change; or by making no apology and no promise to change. Over the course of several rounds of the game, the participant decides whether to entrust money to the partner and also rates their level of trust at the end of the experiment.

In the short term, a promise to change was the most powerful element of the apology—it was more powerful than an apology without it, and most powerful when combined with the other elements of apology. In the long term, trustworthy actions were as effective as promises to change, suggesting that a promise to change signals one's intentions in the short term, such as during an apology conversation. Then, over time, our actions can reinforce this same message (or contradict it). Promises to change make the future look bright. They show the recipient that the apologizer has a concrete plan that seems feasible and . . . promising.

After we make a promise to change, though, it becomes our duty to fulfill that promise—to demonstrate our actual change over time. In the parole hearings, the prisoners who promised to change didn't always follow through—recidivism was all too common. But we can do better. Following through on our promise will take continued introspection, humility, and perhaps asking our partner if they see how we're changing. Apologies, including promises to change, are the first step toward reconciliation, so we might get the chance to actually change and enjoy the rewards of a healthy relationship over time.

In one of Dru and Tashira's final therapy sessions, they describe a conversation that happened at home on Mother's Day. Though their relationship felt like a bumpy roller-coaster ride at the beginning of therapy, over the course of many weeks, they began to enjoy the ride. In Dr. Guralnik's office, they're no longer staring at the ceiling and sighing. They're laughing more. They're sleeping in the same bed now. On that day, though, Dru surprised Tashira with a new T-shirt with the words "Boy Mom" emblazoned in weathered black script across the front. He knew she'd wanted a "Boy Mom" T-shirt for a long time.

"Do you like it?" he asks.

"No. I don't like V-necks, and it's not a good material."

Dru, wounded, says, "You know what, then just throw the shit out." He leaves the room. A few minutes later he returns. "What you just said hurt my feelings."

Hearing Dru's emotional vulnerability, Tashira softens. She apologizes, sincerely, for her harsh reaction. She understands Dru's pain, validates his emotional reaction, and takes responsibility for it. After watching her struggle to apologize for many episodes and therapy sessions, it's incredibly satisfying—for viewers and Dru alike—to watch Tashira make a full apology: there's a direct "I'm so sorry" along with self-castigation ("I was a jerk") and remorse ("I really felt bad").

Then, even better, she follows the apology with a demonstration of change. She puts the T-shirt on. She takes photos of herself in it. She posts them on Facebook. It's not hard to imagine the sense of joy and relief that may have washed over Dru in those moments. In their therapy session, Tashira admits, "When I put the shirt on, I realized I actually liked it a lot." Dru liked it, too. Her apology made him feel much better—about the T-shirt, about his ability to please her, and about her ability to love him back.

They're both laughing about the incident during the session—it feels so silly to have big feelings about a T-shirt. But as we saw with so many difficult moments, this one's not really about the T-shirt—it's about what's happening beneath the earth's crust. Dru and Tashira are finally reciprocating their vulnerability and remorse, first Dru by admitting that his feelings were hurt, then Tashira with a sincere apology about her knee-jerk reaction. Only weeks earlier, their closeness lines were wobbling precariously, poised to permanently part. Through their work with Dr. Guralnik and their caring apologetic behavior, now their closeness lines, like Tashira and Dru themselves, are resting more comfortably together.

Timing

Our recipe for a good apology—one that focuses on our partner's feelings (more than on our own), on the future (rather than the past), and on expressing empathy and contrition (rather than seeking immediate forgiveness)—should also address the question of *when*. When should apologies happen in the context of a conversation or a relationship, and how long after the harm? Many people are reluctant to apologize before they've done a full investigation, understand the problem, and know whether they're really to blame. We're reluctant to apologize if we're not ready to confront our own role in a difficult situation, which often requires internal emotional work.

Still, on the issue of timing, the research is clear: faster apologies are better. Recipients see quick apologies—even those delivered before we understand the transgression itself, or who's at fault—as more sincere.

This presents a conundrum: while a speedy apology is preferable, it's unreasonable to expect the apologizer to quickly accomplish the deep work needed to understand what went wrong—or to come up with an appropriate plan of action to repair it. Understanding one's role in a conflict can take a long time—sometimes years. Luckily, we can issue apologies to acknowledge our partner's suffering immediately, neither shouldering nor dodging blame: "I'm so sorry this has happened. I hate to see you upset. Let's work together to figure out what went wrong."

No matter what happens in the short term, the window of opportunity for apologizing never completely closes. Belated apologies miss out on the power of quick response, but they open the possibility to process issues more deeply and to show actual change rather than just promising to change in the future. They also allow for a bit of emotional detachment from the harm, which can be good—intense, fresh emotions can overwhelm our ability to apologize and to forgive.

After the nose-breaking incident, my toddler Kevin grew. He learned to talk, and his behavior improved. By the time he was three or four years old, he was already labeling his feelings: "I'm mad that you won't let me keep playing." "I'm sad to still be sitting at dinner." Labeling his feelings helped him manage his emotions better—the tantrums faded. I had loved him all along, but his improved behavior made loving him much easier. It was a miraculous relief.

One day when he was seven years old, the two of us were reading together in his bedroom. The main character in his book, Greg Heffley in *Diary of a Wimpy Kid,* was apologizing to his best friend, Rowley. For Greg, as for so many people, apologizing was a rare and courageous act. Observing Greg grapple with the apology, Kevin paused from his reading aloud. He looked up at me and, in a quiet voice, whispered, "Hey, Mom?" I met his eyes. "Remember when I broke your nose?"

"Yeah, I do. Oof! Why do you ask?"

"I'm really sorry."

Five years after the incident, after hearing me retell the story to his dad and to Kevin himself a few times—and after seeing the power of apologies from his beloved protagonist Greg Heffley, he apologized. I was surprised and incredibly moved. Tears filled my eyes once again—this time tears of joy. It had been a mistake, I assured him—it was not his fault. Toddlers, I said, are just learning how to exist in the world, and bigger kids, teenagers, and grown-ups struggle to control and express their frustrations, too. Plus, I said, I thought I looked pretty good with a slightly crooked nose.

He smiled and nodded. Watching him learn to do anything—walking, talking, dribbling a basketball, reading—has been a thrill. But watching him learn to apologize was one of the most rewarding experiences of my life.

Feedback

It can be hard to deliver information that suggests an apology or a change in behavior is needed. But feedback is a *crucial* step in the interpersonal repair process. Whether the glitches to be fixed are small and pesky or deep and hurtful, interpersonal trust depends on our awareness of problems—delivered through many forms of feedback—and our ability to solve those problems, together.

A recent study found that when a survey administrator had food or lipstick smeared on her face, only 3 percent of people gave feedback. And that was just for *crumbs* and *lipstick*—minor pratfalls, not some entrenched bad habit or personality flaw (the horror!). But feedback is a gift—sometimes it takes feeling harsh in the moment to be kind over the long term. And luckily, recent research provides actionable advice about how to give feedback more freely, confidently, and productively.

- **Get the timing right.** Like apologies, feedback is easiest to receive *swiftly,* when both parties can remember concrete details—because explaining those concrete details matters. If you're delivering constructive feedback, it's also important to do so *privately,* not in front of other onlookers, to minimize feelings of embarrassment or shame, and to maximize feelings of safety and trust.
- **State your intentions out loud.** Scientific evidence suggests that stating your good intentions out loud is effective for givers and recipients of dissent alike. If you want to hear constructive feedback, say so explicitly; if you want to give constructive feedback, state your supportive intentions out loud. When receiving feedback, you might remind your partner "I'm a glutton for feedback" or "Don't be afraid to give it to me straight—I want to learn." And when giving it, say, "I want you to thrive" or "I really care about our relationship."
- **Start positive.** The well-known "feedback sandwich," the positive-negative-positive sequence, gets a bad rap because positive and negative ingredients are often smooshed together—the constructive feedback gets lost between squishy, positive bread that feels like perhaps unnecessary sugarcoating. In my research with Leslie John, we've found that the feedback sandwich is actually pretty good, because it's important for positive feedback to come first, no matter *what* comes after it. For constructive feedback to land, it's important to start from a sturdy place of trust. And then it's important to clearly separate positive feedback from constructive feedback—don't smoosh it together in the

same breath, or the same conversational turn. My colleague Frances Frei argues for a 5-to-1 ratio of concrete positive-to-negative feedback: give five sincere, concrete bits of positive feedback before you turn, clearly and honestly, to the constructive stuff. There isn't research on this ratio (yet), but I think it's a great idea. If your relationship is very trusting, then you might be able to dive right into constructive feedback without making your partner feel bad. That's the power of *trust*.

- **Look ahead.** Feedback can be hard to receive because it focuses on things that have already happened—things that can't be retroactively changed. The researchers Mike Yeomans and Ariella Kristal recommend that focusing on forward-looking advice (how the recipient should behave in the future) rather than backward-looking feedback (how the recipient performed in the past) is more likely to be listened to and acted upon effectively.

The Final Assignment

Over the years, I've realized that everyone, even the kindest among us, has a relationship that's gone sideways. A roommate who moved out on bad terms, stranding you with extra rent payments. A high school friend with whom you had one too many quarrels. An in-law who said something rude. A sibling who didn't attend a funeral. A partner lost to a bad breakup. Or a friend who simply drifted out of your social orbit. Whatever the reason, people disappear from our lives. Our closeness lines go in different directions, to different pages, to different worlds.

What happens when you want to reconnect? I'm intrigued by the conversations that transpire after long periods of broken contact. The process of repairing what was broken, and what seemed like it would remain quiet forever, is difficult and moving. People in these situations may have lost all trust for one another—they've long ago missed the window for repairing trust through everyday apologies. For whatever reason, they've let their relationship die. Restoring lost trust may require an extraordinary apology from both parties, and a prolonged reckoning with what went wrong.

For their final project in the TALK class at HBS, I ask my students to have a conversation, record it, and reflect about the experience using the skills they've learned. I give them a suite of options—different conversational challenges—so that they can select a project that feels personally meaningful. One of the options is called Relationship Reboot, in which they reach out to someone they've lost touch with and refresh their relationship. Other, less daunting options are available, like expressing gratitude to someone or creating a podcast episode. So only a few students choose the Relationship Reboot each semester, about five in every section of ninety students. The Relationship Reboot projects are often some of the most fascinating of the semester. But there's one in particular that I still think about—a project submitted by a student named Dev.

Dev was an exceptional teenager—calm, kind, smart, hardworking, and a bit socially anxious. He was the kind of kid many parents hope for and teachers adore. He was thriving in eleventh grade, and he had his sights set on top colleges. That was when he met Anil, a seventh-grader who reminded Dev a lot of himself. They were both from Indian families who had immigrated to the same suburban community outside Sacramento, California. Anil wasn't as confident as Dev, but he wanted to be. He was a good kid, and he idolized Dev's success. They bonded quickly, and Anil became close with Dev's family and friends, too. A standout student, Dev was determined to help Anil thrive both personally and professionally. He was an excellent mentor for Anil, serving as a close ally through Anil's middle school, then high school—they were like brothers. Dev called him often to check in. Even from a distance, Dev helped Anil sort out his college applications and made sure they looked good before Anil sent them off.

In 2018, Anil graduated from the University of Southern California. Struggling to find a job, he reached out to Dev for help. Dev had been climbing the ranks at a financial services start-up and was thrilled to help Anil land an entry-level financial analyst position at the same firm. He'd worked hard to attain early professional success so that he might uncover opportunities to help underrepresented minorities, especially Anil, whom he cared about so much.

But Dev made it clear to Anil that it would be harder for him to keep the job than it had been to get it. In fact, succeeding in the role would require a lot of work to get up to speed, especially mastering technical skills in Excel and other analytical tools used at the firm. Alas, Anil fell behind on his work almost immediately. He often seemed confused and overwhelmed. Over many months, he didn't develop the technical skills required for the job—at least not fast enough to fulfill his responsibilities. His struggles made Dev feel angry and disappointed—embarrassed that he had stuck his neck out for Anil, who seemed like he wasn't even trying.

In a last-ditch effort to help, Dev cleared Anil's plate of all other work responsibilities except learning Excel. Anil had six weeks to develop a case study and thereby showcase his technical competence so that he could get back to work. After the intensive training period, Anil failed to complete his task. Dev was devastated—and exasperated. He had no choice but to terminate Anil's employment, driving a heartbreaking wedge between them (and making relations between their mutual friends and families tense).

For two years following Anil's firing, the two didn't speak—they lost touch. Dev felt the situation was crystal clear: he had given Anil an opportunity, and Anil had let him down. He was hurt, angry, and disappointed. Occasionally, Anil reached out—over text, email, and phone—but Dev didn't respond. Over time, though, Dev started to feel bad about it.

The idea of talking to Anil for the first time in two years made Dev nervous. To prepare for the call, he leaned on tactics he'd learned in the course, starting with a list of topics. He knew he needed to probe Anil's perspective and offer his own on what had happened, so he created a list of very specific questions. And he thought about how to inject some levity into the conversation, to cut the tension. He'd start with some small talk, working up to the conversation about Anil's firing. Though it made him anxious, Dev knew this was a conversation they needed to have.

On a Wednesday in April, Dev called Anil, who picked up after just one ring. Heart racing, Dev said hi but immediately sensed the sadness and concern in Anil's voice. Dev tried for small talk, asking Anil how he'd been doing. But the elephant in the room was too colossal to ignore.

Anil said, "We used to be so close, and you were like my big brother when things ended. It was like I didn't exist in your life anymore. It's been pretty crappy to be honest." Anil's emotions were close to the surface. Dev heard Anil's voice quiver.

"I know. I've been thinking about our friendship and how things ended and wanted to chat with you about it. I'm sure there are two sides to the story, and I recently came to the realization that we need to talk and figure out a path forward, if there is one."

Anil admitted, "There were some moments and things I definitely screwed up, but I think there is some responsibility you need to take as well."

"Interesting," Dev said. "Mind saying more?"

Anil described his feelings of abandonment: "I remember I promised to work hard night and day, but I made that promise thinking you would be there to help me, and I guess I just felt very, very abandoned."

"Why," Dev asked, "didn't you address these topics directly with me back then?"

"I was scared," Anil admitted. "Very scared that I let down my big brother. Scared that I wouldn't be able to ramp up quickly and do what I needed to do."

They took turns explaining their memory of how things unfolded at the firm. Dev was disappointed in Anil's performance, and Anil felt abandoned by Dev, who left him to sink or swim on his own. He wanted to live up to Dev's high standards, but the job was quite difficult for someone with no experience with financial modeling. He expected Dev to coach him more and to join him in meetings as an ally. When Dev didn't live up to this coaching role, Anil struggled. He didn't sleep or eat well. He was often sick and knew he was letting Dev down.

Listening intently, Dev responded without a hint of defensiveness. "I want to apologize for making you feel alone and not supported," he said. "In fact, I wanted the opposite of that and to provide you an opportunity to really grow." Dev had learned on the job by asking questions and working autonomously, and he assumed Anil would want to do the same. But then he realized that everyone is different, and he owed Anil an apology: "I am so sorry, Anil."

Anil was taken aback by Dev's heartfelt apology. "Really, you made my year already," he said. "I'm sorry for not giving it my all and being a terrible communicator."

"You're not the only one. I'm so glad we're connected again."

A couple days later, they got back on the phone. They opened with so many apologies that Dev suggested a rule: No more saying sorry—a sweet joke that acknowledged their mutual apologies and tried to make light of the situation. "We both messed up and owned up to it," Dev said. "I think of you as my little brother and want to get back to the phase where we are acting like that."

During the remainder of the conversation, even though neither man used the word *sorry,* both admitted their failings in the past, expressed remorse, and looked hopefully to the future. Like old friends, they moved toward small talk and life updates. They made a plan to watch a Los Angeles Lakers basketball game together on Zoom a few days later. They weren't just dreaming about being close again—they were doing it.

Conversing with Less Fear, More Excitement

The ballad of Dev and Anil shows the power of reciprocity. As soon as Dev became responsive and vulnerable, admitted his mistakes, and apologized, Anil did the same. Dev's warmth and courage kicked them into an upward spiral that not only repaired their rift but also set them on a positive trajectory for the future. Their story shows that belated apologies can be powerful—it may not be too late to recover lost trust, even when it feels hopeless. The passage of time can give us an opportunity to gain a new perspective, emotional distance, and the space to show real change.

The TALK maxims can always help us, but they're especially powerful in poignant conversations like Dev and Anil's reboot. Topic forethought helped Dev prepare not only logistically but emotionally. Even if Dev didn't raise the "very specific questions" he had written down ahead of time, the act of writing them down made him more confident and level-headed when he was met by Anil's charged emotions.

Several times during their conversations, Dev asked Anil excellent questions. He asked if Anil might be open to sharing his perspective—to "figure out a path forward if there is one." He asked follow-up questions ("Interesting. Mind saying more?"), and they were very effective. With just four words, Dev showed his responsive listening, validation, and care—leaving a gracious opening for Anil to give his perspective.

Dev went on to ask several other questions like "Do you mind if we talk about that a little?" (topic steering), "So what happened there?" (follow-up), and "Did you know what position you put me in?" (perspective-getting). It's an impressive show of Dev's question-asking skills.

Their first conversation was emotional, yet they still found opportunities for levity. In particular, they poked fun at the fact that Dev had needed a professor to nudge him to make the call—and at the awkwardness of recording it. The mood in their second chat was lighter overall, with Dev's joke that apologies weren't allowed anymore, followed by the shift to lighter topics. They talked about betting on the Lakers and debated whether their siblings would be giving their parents more grandchildren. These subtle moves helped to keep things pleasant, despite the heaviness of their main topic.

Of course, the core of Dev's conversational success lay in his kindness. It took tremendous courage to call Anil, and when he did, he used respectful language—when it would have been easy to make a point about Anil's fumbling professional experience in a way that he would find insulting. With responsive listening and affirmations like "Interesting. Mind saying more?" and "I understand. I'm so sorry," Dev's conversational choices are a master class in focusing on his partner, Anil, and validating his feelings. Kindness was the main mission of the project, and he nailed it.

Though Dev began the semester nervous in social situations and somewhat convinced he was better off spending more time alone, in his final project, he showed a change of heart. He said he wouldn't have reconnected with Anil without what he had learned in the course. He felt transformed: learning to use TALK helped him be a better friend and a better person. His gratitude wasn't just for his renewed connection with Anil but for all the relationships in his life yet to come. "For the first time," he wrote, "I'm truly excited to connect with people."

THREE KEY TAKEAWAYS FROM CHAPTER 8

Apologies

- **Apologies are remarkably powerful.**
- **Apologize frequently** and sincerely—don't make it about yourself.
- **Promise to change,** then do what you've promised.

Epilogue

A FEW YEARS AGO, I visited a dear friend whom I had not seen in years. I'd moved to a new city, we'd both had children, and she'd been in chemotherapy. To help us catch up, she created a topic list. And she even went a step further, labeling her topics with Whitney Houston song titles:

- "I Wanna Dance with Somebody"—Tell me about the last time you went dancing!
- "Higher Love"—I took my baby on an airplane for the first time. Let me share the horrors.
- "Run to You"—Can we talk about your running regimen and what it takes to train for a marathon?

We laughed and sang and hugged and gossiped and reminisced and cried. It was one of the best and most meaningful conversations of my life.

Even great conversations, though, have moments that are stilted. In that wonderful conversation with my friend, some topics became stale, our energy flagged, her mouth became parched, I felt overcome with a nostalgic sadness that I wasn't sure I should express to her. But somehow those moments didn't make the conversation any less great. *Why?*

Seeking Richness

Until 2021, social scientists had focused on two factors that comprise human well-being: happiness (feeling good) and meaning (feeling that your life serves a purpose). But something was missing from this age-old formulation. In 2020, psychologists Shigehiro Oishi and Erin Westgate confirmed a previously overlooked third pillar of well-being: *psychological richness*—a life characterized by variety. While a person with a steady and rewarding job and marriage may have a happy and meaningful life, it may not be rich in diverse, fascinating, new, or exciting experiences.

The key point here is that a happy and meaningful life can also be *boring*. Our need for psychological richness is why your steadily employed friends might choose to frequent art galleries, try drugs, play on a hockey team, take guitar lessons, enroll in an executive education course, take up scuba diving, or read a book like this one—to make their minds spark and ping in new and exciting ways. It's also why aiming for constant conversational greatness isn't the right goal. Psychological richness suggests that the ups and downs of life, the bitter and the sweet—a wide-ranging distribution of experiences—are necessary, good, and, perhaps most important, unavoidable.

Perhaps it is no surprise, then, that talking to a mix of people, for different purposes, on a far-ranging mix of topics, makes us feel the most alive. In fact, a key finding from my own research is that connecting with a diverse set of conversation partners—new friends and old, lovers and new acquaintances, colleagues and strangers—makes people happier than speaking with a narrower range of partners. Diverse conversations with a wide range of people and relationship types inspire us to feel diverse emotions—happy, sad, anxious, frustrated, calm, excited, devastated. And this diversity of feelings is what, ultimately, makes us happy.

Are you still feeling nervous about conversation? It's okay! You're human! As I remind my students, just getting out there and practicing is all that's required. And luckily, we get lots of opportunities to practice. Those in pursuit of better conversation should not be looking for *enormous* changes. Can you ask one more follow-up question than you normally would? Can you prepare one exciting topic? Can you make your partner laugh at least once? Can you use one call-back to show you're listening? These small moves won't always help, but more often than not, they *will*. And small improvements, across a lot of diverse conversations, make a big difference. To get a little better at conversation—or even to just *want* to get better—is to be more fully human.

If you keep the TALK maxims in mind, you'll see that you can cast some common worries aside. You needn't worry about what to talk about—you can prep topics! You needn't worry about muttering or stumbling—lean into repair strategies! You needn't worry about awkward silences—kick-start a new topic! You needn't worry about impressing people with your knowledge or expertise—just listen and ask questions! You needn't worry about sensitive topics or hostile conflict—stay civil, be receptive! You needn't worry about when to leave—just go! In short, you got this.

And I'll let you in on a little secret: you're already better at talking than you think. In 2023, psychologists Christopher Welker, Jesse Walker, Erica Boothby, and Tom Gilovich examined people's assessments of their own conversational ability. They found that we are overly pessimistic about our skills in conversation compared to other common activities like driving, reading, and running. On most tasks, people tend to overestimate their abilities compared to others' (the well-known "above-average effect"), but conversation is the only task on which people tend to show a "below-average effect." We fixate on the moments of awkwardness and difficulty we experience, and we tend to think they're our own fault. While we view others as embodying the Myth of Naturalness, assuming that their charisma is intrinsic and effortless, we view ourselves as embodying the Myth of Suckiness, ruminating about all the ways we've screwed up, despite our best efforts to the contrary. But conversation is co-constructed, it can't be scripted, and even the best talkers are far from perfect. We're all better off celebrating when *anything* goes well rather than ruminating about all the things that (predictably) don't.

So you're already better than you think. And now you have tools to become even better. In his novel *East of Eden,* John Steinbeck wrote, "Now that you don't have to be perfect, you can be good." Once we give ourselves and everyone else permission to flop, that's when our conversations and relationships can get *really* good—happy, meaningful, and psychologically rich.

Flouting and Making It Stick

Here's another little secret: the TALK maxims aren't really maxims. *What?!!* It's true. Real maxims are rules of conduct summarized in short, pithy statements. While real maxims can be measurably adhered to, they can also be *flouted.* Paul Grice intended for his maxims (be truthful, concise, relevant, and clear) to be both followed and flouted—for his rules to, sometimes, be broken. And he believed that when everyone is in on the flouting—when we break the rules on purpose—that's when conversation gets really good.

If not rules of conduct, then what are the TALK maxims? They're *reminders.* Reminders to do the things we want to do—to be our best selves and to help others be their best selves, too. We need these reminders because we're likely to forget, overlook, or abandon our intentions when we get lost in the sauce of conversation.

At the end of the TALK course, my students ask: How can I make the TALK maxims stick? How can I be sure I don't forget these fundamental reminders? As a parting gift, I give them silicone bracelets emblazoned with "Topics-Asking-Levity-Kindness"—a portable reminder of the TALK maxims. Something like this may help you, too. But truly, I think the best answer is simply *practice*. Talk about TALK to others. Teach it. Live it. Reread it. Reflect about it. Ask a friend or partner or colleague to give you feedback about how it's going. Record and listen back to a conversation every once in a while. See what works for you, and what doesn't. What are you doing well, and where are your wobbles? Like all effortful things that can become less-effortful habits over time, I hope your TALK skills will stick.

Light Up the World

Remember Olivia de Recat's closeness lines, tracing the trajectories of our relationships? What if we think of our closeness lines and the conversational knots spread out along them, instead, as a string of *lights*? Every knot on every string has the potential to become one of those festive white café lights that make backyard patios and al fresco dining by night so thrilling. What is it that makes some lights twinkle and burn bright, when others flicker and burn out? In this book, we've untangled many filaments and traced the sources of interpersonal electricity. We've seen how we might try to avoid tepid brown-outs and repair blown fuses. Now it's time to pull the filaments back together.

Erving Goffman once said that conversation "is the spark, not the more obvious kinds of love, that lights up the world." As we gaze at the great web of lights and strings all around us, what if we endeavor to make each bulb burn a little brighter? There will be moments when our conversations threaten to flicker or burn out, but there will also be moments of brilliance. Let's get out there and light up the world, one conversation at a time.

APPENDIX

Putting TALK into Practice

This appendix offers two different ways to put your TALK skills into practice. First, I list several reflection exercises you can use on your own. Second, I include several conversation exercises to engage in with others.

Reflect

These exercises will nudge you to reflect privately about your conversational lives.

Increasing awareness of your goals. Before your next conversation, plot your goals on the conversational compass. You may have many, or only a couple. Remember, you always have at least one, even if it's simply to pass the time, be polite, or have fun. Then write down your predictions about your partner's goals. Reflecting on and codifying what you're trying to accomplish helps to clarify and guide how you should behave, what conversational success might look like, and how you can think about measuring it after the fact.

Your personal history of conversation. Just as the meaning of conversation has evolved over time and place, your personal view of conversation has evolved over the course of your life. Can you identify different phases of your development and how you thought differently about conversation—or practiced it differently? What events and/or people precipitated your personal conversational evolution?

Coordination problems. Think back to a recent conversation. Did you and a partner fail to coordinate around shared understanding—a moment when it became obvious that you weren't understanding each other's beliefs, preferences, or knowledge? (The answer is almost certainly yes, in some way.) What went awry? And how do you know? What if you called that person now and tried to re-coordinate? Could you fix the crack in your shared reality?

Flouting Grice's maxims. Think back to a conversation you just had. Were you and your partner perfectly truthful, concise, relevant, and clear? When? Why or why not? Do you think your rule-breaking was intentional or accidental? Was it good or bad?

Sources of difficulty. Think about your last difficult conversation, one that had at least some parts (or a single part) that made you feel confused, sad, mad, anxious, or frustrated. What triggered those feelings? What was it about that interaction that was hard? How did you behave then? How would you behave if you could do it again?

Cooperative or competitive? Many conversational choices don't work the same way when we are agreeing or disagreeing—cooperating or competing. Flattery, sarcasm, topic changes, probing questions: What strategies have you noticed work well with close others during cooperative moments, but flop (or incite problems) during more difficult moments or with more difficult partners?

Identity map. First write down your identities in the following categories: family, work, social/cultural/biological, schooling, hobbies, and other. (Mine would be mother, wife, daughter, twin, professor, behavioral scientist, mentor, friend, singer, basketball player, Pilates devotee, chronic laugher, white, woman, Irish heritage, American, raised Catholic, etc.) Then plot those identities on a spectrum from unobservable to observable. What aspects of your life are invisible until you talk about them? To whom? Are those identities associated with different levels of power or status? How do they influence how you communicate?

Digital communication audit. Write down all the incoming and outgoing messages, across all modes of communication (text, email, phone, video call, face to face, etc.), over the course of twenty to thirty busy talk minutes of your life. What surprised you about the amount and sequencing of the messages? Who did you respond to fastest? Who did you ignore? Why? What questions does this audit raise for you?

Your digital identity. Go ahead, Google yourself. Who are you, digitally? If a stranger trolled you on Facebook, LinkedIn, Instagram, or Snapchat, or emailed or texted with you before meeting face to face, would they get to know who you really are? Why or why not? What aspects of you would surprise them if you met face to face? Is your face-to-face conversational style knowable through digital channels? Is it markedly different from your digital communication style? Are you setting people up for a disappointing letdown or a delightful surprise, or are you cultivating accurate expectations?

Your most meaningful conversations. What were some of the most meaningful conversations of your life? Can you make a list? Write them down? Describe them as stories? What details do you remember? What made them special or important?

Practice

These exercises will give you ways to workshop and practice your TALK skills while engaging in conversation with others.

Pick a goal, any goal. Pick one very clear goal for your next conversation. Goals like the following might work well: *Learn as much as you* can. *Make it fun for your conversation partner. Don't let your mind wander. Be radically honest (not even little white lies). Be as likable as possible. Make lots of jokes.* If you're brave enough to try, you'll likely be surprised by how tough it is to keep your singular purpose in mind without getting distracted or drifting into old habits, such as, in addition to learning as much as you can, profoundly caring that people like you.

Prep topics. Take thirty seconds to write down five possible topics for your next conversation. When the time comes, you don't *have* to have the topic list in front of you (although it can help)—simply writing down the topics now (or adding them to your digital calendar notes) will help you remember them on the fly. You don't have to raise any of these topics during the conversation, but enjoy the comfort and confidence of knowing that you *can*.

Be a topic follower. In an upcoming conversation, try to let your partner(s) take the lead. Give them space to choose whether to switch or stay on topics. This may be a challenging exercise in self-restraint, especially for dominant conversationalists.

Be a topic leader. In an upcoming conversation, try to take the lead through assertive topic management. If a topic seems juicy (rich, fun), ask follow-up questions to stay on them. If the juice seems to be running dry (if you notice uncomfortable laughter, longer mutual pauses, or redundant statements), switch assertively to a new topic. You can always call back to topics that didn't seem to get enough airtime. This may be a challenging exercise, especially for quiet, hesitant, or polite conversationalists—or those in the habit of staying on topics for long stretches of time.

Asking and answering hard questions. Make four lists of questions: questions you'd be nervous to *answer* (1) at a dinner party and

(2) during a job interview, and questions you'd be nervous to *ask* (3) at a dinner party and (4) during a job interview. Sit down with a trusted partner and role-play asking and answering these questions.

Never-ending follow-ups. In an upcoming conversation, try to ask follow-up questions until the conversation ends. You should respond to whatever your conversation partner says, and it's okay to switch topics, too, but always end each speaking turn by asking a follow-up question. Did your partner notice? How did it go? Many people find this to be a surprisingly easy way to ascend the topic pyramid.

Be a journalist. In an upcoming conversation, channel your inner Oprah (or Rachel Maddow, or Anderson Cooper). Try to learn as much as you can about your conversation partner *without making any assumptions* about them. Try to avoid interjecting your own views or stories. Practice balancing follow-up questions and topic-switching questions.

Escalate. The next time you notice that someone seems quiet, sad, or low energy, focus your energy on improving their mood, maybe even on making them smile, with a tiny injection of humor or warmth. Boredom is a quiet killer of conversation. If you spot it, try to fix it!

De-escalate. The next time you witness or are engaged in a heated debate, shift your motives to focus 100 percent on de-escalation. Completely give up on being right, and instead focus on making things right, mood-wise. What happens?

Levity reframe. Tell someone about a difficult experience in your life, recounting as many details as possible. Then tell the story again, but try to convey it with as much humor and mirth as you can muster. Can you do it? What did you notice about the two different versions? Another version of this exercise: Ask someone about their work, then ask them what they're passionate about or love doing. Notice how their energy changes.

Give a compliment. Make a point to give the next person you see a sincere compliment—something you like or admire about them. It will light up their world more than you think.

Validation challenge. At the beginning of each statement, every time you talk, start by affirming what your partner just said: "I love how you said that," "Oh, how interesting," "What a great perspective," "Thank you for asking—how great," "It's so interesting that you said . . ." "It makes sense that you felt that way." Try to focus on validating their feelings, even if you go on to disagree with their opinion or belief.

Practice paraphrasing. In your next conversation, especially if it's at work and/or in a group, try paraphrasing what has been previously said by one, two, or three conversation partners. Can you tie what they've said together in a logical way? Can you summarize across multiple people? Are they agreeing or disagreeing? Can you draw a pithy distinction between what one person has said versus another? You might start with the phrase "What I'm hearing here is . . ." or "Am I understanding correctly that . . ." or "Just to make sure we're on the same page, I think what you're saying is . . ."

Limited listening. During your next one-on-one video call, spend some time with your video turned on, and some time with the video turned off. How do you convey that you're listening when the camera is on? What about when the camera's off? How does your behavior change? Or in your next face-to-face conversation, try having everyone close their eyes for part of it. How does your listening change? Are you able to coordinate turn-taking? In what ways are you listening more or less attentively?

The power of gaze. Stare into someone's eyes for four minutes without talking. This is said to bring people closer together. Can you do it? How did it feel?

Companionable silence. Do you have people in your life with whom you can enjoy companionable silence? Who are they? How long have you known them? Are there people with whom shared silence would probably be okay, but you haven't tried it yet? In your next encounter with those folks, give it a try. Shared silence can bring you closer, quickly.

Vipassana (silence retreat). I know it sounds crazy (really, I thought it was crazy), but how long can you do a silence retreat (i.e., stay silent in the presence of others)? Can you make it through a whole conversation? A whole day? What about a week? What do you notice?

Feedback challenge. Seek constructive feedback (that you know you need) from two people you trust to give it. You should seek feedback on things that will help you in your work and life. Encourage them to be direct and honest, especially on things you might be able to improve. If you start to feel defensive, they are likely touching on a real area of opportunity. Remember, feedback is a gift, especially constructive feedback.

Relationship reboot. Think of someone you care about whom you've lost touch with or have a relationship with that's hurting. Reach out to them to reconnect and/or reboot your relationship. Use your TALK skills to entertain them, listen to them, and seek to understand their perspective.

Apologize. Leverage your apology skills to apologize, sincerely, to someone you care about. You'll be glad you did.

Teach TALK to others. Engage people in a conversation about what you've learned in this book. (Please cover other topics during those conversations as well. Good conversations don't linger too long on topics that have grown stale!)

Conversation masters. Who do you regard as particularly good (or bad) at conversation? Give them a call—see what you notice about them after reading this book!

Words to Live By

What's your favorite quote about communication? Is there a famous phrase or saying that captures your conversational values well? Did anything in this book remind you of that quote? Quotes are appealing because they can help us summarize our understanding and beliefs. Here are some of my favorites:

"Jazz is a conversation, but a nuanced, swift, and complicated one."
—WYNTON MARSALIS

"You've got to learn your instrument, then, you practice, practice, practice. And then, when you finally get up there on the bandstand, forget all that and just wail." —CHARLIE PARKER

"A lack of seriousness has led to all sorts of wonderful insights."
—KURT VONNEGUT

"Play is often talked about as if it were a relief from serious learning. But for children, play *is* serious learning." —FRED ROGERS

"We were each other's ideal audience; nothing, not the slightest innuendo or the subtlest shade of meaning, was lost between us. A joke which, if I had been speaking to a stranger, would have taken five minutes to lead up to and elaborate and explain, could be conveyed . . . by the faintest hint. . . . Our conversation would have been hardly intelligible to anyone who had happened to overhear it; it was a rigamarole of private slang, deliberate misquotations, bad puns, bits of parody, and preparatory school smut."

—CHRISTOPHER ISHERWOOD

"A high school student wrote to ask, 'What was the greatest event in American history?' I can't say. However, I suspect that like so many 'great' events, it was something very simple and very quiet with little or no fanfare (such as someone forgiving someone else for a deep hurt that eventually changed the course of history). The *really* important 'great' things are never center stage of life's dramas; they're always in the wings."

—FRED ROGERS

"When true silence falls, we are still left with echo but are nearer nakedness. One way of looking at speech is to say that it is a constant stratagem to cover nakedness."

—HAROLD PINTER

"Every difficult conversation starts with a sentence."

—SELINA MEYER, *Veep*

Some Topics to Try

These are "36 questions that lead to love" from Arthur Aron's research:

1. Given the choice of anyone in the world, whom would you want as a dinner guest?
2. Would you like to be famous? In what way?
3. Before making a telephone call, do you ever rehearse what you are going to say? Why?
4. What would constitute a "perfect" day for you?

5. When did you last sing to yourself? To someone else?
6. If you were able to live to the age of ninety and retain either the mind or the body of a thirty-year-old for the last sixty years of your life, which would you want?
7. Do you have a secret hunch about how you will die?
8. Name three things you and your partner appear to have in common.
9. For what in your life do you feel most grateful?
10. If you could change anything about the way you were raised, what would it be?
11. Take four minutes and tell your partner your life story in as much detail as possible.
12. If you could wake up tomorrow having gained any one quality or ability, what would it be?
13. If a crystal ball could tell you the truth about yourself, your life, the future or anything else, what would you want to know?
14. Is there something that you've dreamed of doing for a long time? Why haven't you done it?
15. What is the greatest accomplishment of your life?
16. What do you value most in a friendship?
17. What is your most treasured memory?
18. What is your most terrible memory?
19. If you knew that in one year you would die suddenly, would you change anything about the way you are now living? Why?
20. What does friendship mean to you?
21. What roles do love and affection play in your life?
22. Alternate sharing something you consider a positive characteristic of your partner. Share a total of five items.
23. How close and warm is your family? Do you feel your childhood was happier than most other people's?
24. How do you feel about your relationship with your mother?

25. Make three true "we" statements each. For instance, "We are both in this room feeling . . ."
26. Complete this sentence: "I wish I had someone with whom I could share . . ."
27. If you were going to become a close friend with your partner, please share what would be important for him or her to know.
28. Tell your partner what you like about them; be very honest this time, saying things that you might not say to someone you've just met.
29. Share with your partner an embarrassing moment in your life.
30. When did you last cry in front of another person? By yourself?
31. Tell your partner something that you like about them already.
32. What, if anything, is too serious to be joked about?
33. If you were to die this evening with no opportunity to communicate with anyone, what would you most regret not having told someone? Why haven't you told them yet?
34. Your house, containing everything you own, catches fire. After saving your loved ones and pets, you have time to safely make a final dash to save any one item. What would it be? Why?
35. Of all the people in your family, whose death would you find most disturbing? Why?
36. Share a personal problem and ask your partner's advice on how he or she might handle it. Also, ask your partner to reflect back to you how you seem to be feeling about the problem you have chosen.

This list covers fifty topics that you could raise with anyone. They are drawn from my research with Mike Yeomans (and some of them overlap with the "36 questions that lead to love"):

1. What do you do for work? What do you like about it?
2. What do you enjoy doing in your free time?

3. Did you do any sports or clubs in high school?
4. Why do you participate in online studies?
5. Do you have any plans for the weekend?
6. What's something random about you?
7. Have you read anything interesting recently?
8. Have you tried anything new recently that was particularly fun?
9. Are you a religious person? Why?
10. What games have you played in the past that are most memorable?
11. What is your favorite kind of music?
12. How do you most enjoy spending time with your family?
13. Do you have any fruit trees, plants, or a garden?
14. What's your favorite movie?
15. Do you have any plans for the rest of the day?
16. Do you like where you live or do you want to move?
17. Do you travel much?
18. What do you enjoy doing when the weather is beautiful?
19. Do you have a favorite type of food?
20. For what in your life do you feel most grateful? Why?
21. What was an embarrassing moment in your life?
22. What's the strangest thing about where you grew up?
23. Who is the luckiest person you know? Why?
24. If you could teleport by blinking your eyes, where would you go right now?
25. What is the last professional sports game or match you watched?
26. What is the last concert you attended? Why?
27. If you had to perform music in front of a crowd, what would you do?
28. What TV show have you watched recently?
29. What is the cutest thing you've seen a baby or child do?
30. Given the choice of anyone in the world, whom would you want as a dinner guest?

31. Would you like to be famous? In what way?
32. Before making a telephone call, do you ever rehearse what you are going to say? Why?
33. What would constitute a "perfect" day for you?
34. When did you last sing to yourself? To someone else?
35. If you were able to live to age of ninety and retain either the mind or the body of a thirty-year-old for the last sixty years of your life, which would you want?
36. Do you have a secret hunch about how you will die?
37. If you could change anything about the way you were raised, what would it be?
38. If you could wake up tomorrow having gained any one quality or ability, what would it be?
39. Is there something that you've dreamed of doing for a long time? Why haven't you done it?
40. What is the greatest accomplishment of your life?
41. What do you value most in a friendship?
42. If you knew that in one year you would die suddenly, would you change anything about the way you are now living? Why?
43. What does friendship mean to you?
44. How close and warm is your family? Do you feel your childhood was happier than most other people's?
45. How do you feel about your relationship with your mother?
46. When did you last cry in front of another person? By yourself?
47. What, if anything, is too serious to be joked about?
48. If you were to die this evening with no opportunity to communicate with anyone, what would you most regret not having told someone? Why haven't you told them yet?
49. Your house, containing everything you own, catches fire. After saving your loved ones and pets, you have time to safely make a final dash to save any one item, what would it be? Why?
50. Of all the people in your family, whose death would you find most disturbing? Why?

The following are ten of my go-to topics to raise with strangers. These questions I like to ask can help you move up the topic pyramid or kick-start a really great conversation:

1. What are you excited about lately?
2. What is something you're good at but don't like doing?
3. What's something you're bad at but love to do?
4. Is there something you'd like to learn more about?
5. Is there something you'd like to learn how to do?
6. What can we celebrate about you?
7. Has someone made you laugh recently? What happened?
8. What is something cute your {kid/friend/pet/partner} has been doing recently?
9. Did you grow up in a city?
10. Have you fallen in love with any new {music/books/movies/shows} lately?

ACKNOWLEDGMENTS

I am deeply grateful for the community of people that helped me write this book, starting with a giant thank-you to my book team, without whom *Talk* would not exist. Alison MacKeen, you are the most perfect agent—colleague, cheerleader, sounding board, mentor, and friend—I could have wished for. Thank you for understanding me and my book from the get-go and enduring so many conversations along the way. To my insightful and blessedly funny editor, Kevin Doughten, despite your go-to joke that you might be the "worst person to edit this book," you were in fact the best person to edit this book. Thank you for understanding my vision and persistently mixing it into a delicious cocktail, along with your amazing team at Crown: Penny Simon, Chantelle Walker, Rachel Rodriguez, Amy Li, and Jess Scott. Thank you for believing in *Talk* and bringing it into the world. To my caring and expert publicist, Aileen Boyle, thank you for helping *Talk* reach more people—and for not making me spend too much time on social media. To my fabulous team at Verto Literary—Gareth Cook, Eli Mennerick, Kate Rodemann—thank you for teaching me to chase the energy, cut the chaff, and make it magical. You made me feel like Xena, the warrior princess, in this Herculean task. To my fastidious and loving research assistants, Katie Boland and Taqua Elleithy, thank you for providing moral support, lending expert assistance reading, brainstorming, compiling references, and helping me laugh every day. To my freelance editorial helper, Fiona Furnari, and my fastidious fact-checker, Hilary McClellen, thank you for bringing your enthusiasm, gratitude, and talent to bear on this book. To my wonderful babysitters, who have cared lovingly for my children so I could focus on writing over the last five years—Samantha Ford, Sarah Ford, Riley O'Connell, Clara Cox, Victoria Cox, Katherine Timmons, Nicole Bonaccolto, Whitney Byers, Amanda McGinnis, Meagan Stella, Molly Naser, and many other talented teachers—thank you from the bottom of my heart.

I am also deeply grateful to my research collaborators. Who knew that fifteen years' worth of conversations about conversation could be so exciting and meaningful? I admire each of you for your unique intellect, dedication, and creativity, and, most of all, for your patience and love: Maurice Schweitzer, Mike Yeomans, Hanne Collins, Nicole Abi-Esber, Katie Boland, Karen Huang, Mike Norton, Leslie John, Todd Rogers, Ryan Buell, Ting Zhang, Grant Donnelly, Julia Minson, Ariella Kristal, Ovul Sezer, David Levari, Gus Cooney, Adam Mastroianni, Emily Prinsloo, Ethan Burris, Jonah Berger, Brad Bitterly, Matteo Di Stasi, Jordi Quoidbach, Jimin Nam, Maya Balakrishnan, Adam Galinsky, Julian De Freitas, Serena Hagerty, Catarina Fernandes, Jeremy Yip, Emma Levine, Hayley Blunden, Jennifer Logg, Laura Huang, Brian Hall, Lizzie Baily Wolf, Julia Lee, Sunita Sah, Tami Kim, Hengchen Dai, Ethan Kross, and Daniel Gilbert. Thanks are also owed for prebook-era research assistance from Ethan Ludwin-Peery, Holly Howe, and Trevor Spelman—thank you for your attentive and dedicated work.

I am so appreciative of the Harvard Business School MBA students and executives who have taken my course. Thank you for joining me on your personal journey toward "gooder" conversation and connection. It's an intimate journey that requires vulnerability and openness, and I am so grateful you chose me to be your guide. A special thank-you to those whose stories have stuck in my mind and heart, so that I could share some of them here. To those who have helped me develop the course—Mike Norton, Naomi Bagdonas, Matt Weinzierl, Kristin Mugford, Rachel Greenwald, Ashley Martin, Anthony Veneziale, Kathleen McGinn, Tom Gilovich, Gus Cooney, Oliver Badenhorst, Irene Kwok, Wallace Lukens—and those who have believed enough in TALK to teach it in other places—Matt Abrahams, Naomi Bagdonas, Jennifer Aaker, Maya Rossignac-Milon, Erica Boothby, Maurice Schweitzer, and Shimul Melwani—thank you. And to my course visitors, Amit Bendov, Elise Keith, Anthony Veneziale, Rachel Greenwald, Naomi Bagdonas, Michael Lewis, Mike Norton, Dani Klein Modisett, and Todd Rogers, thank you for lending your wisdom and care to me and my students.

Most of all, my heart bursts with gratitude for my friends and family—for all the things you do to help me every day. To my friends, what is life without you? Thank you for all the conversations over texts, meals, cocktails, walks, and youth sports games. To my bandmates in The Lights—Bishop Levesque, Ryan Buell, Mike Norton, and Derek Brooks—I'm so grateful we get to coordinate as a group. Making music together is sometimes hard and sometimes easy but always miraculous. Most of all, I am grateful for my family—a winning lottery ticket I didn't earn, but appreciate deeply. Mom and Dad, thanks for telling me I was a good writer. More importantly, thanks for telling me to take a break once in a while, go for a swim, hang out with friends, and eat a banana. Sarah, thank you for giving me the most joyful life as a twin. Being your sister has shown me the power of mutual understanding, and has inspired me to spend the rest of my life trying to help others find it, too. Brendan, thank you for tolerating me and Sarah. I'm sorry no one could make us stop talking (or singing). You're the most patient of all. To my in-laws, thank you for believing me worthy of marrying Derek and making me feel loved like one of your own. To my children, Kevin, Grady, and Charlotte, I will never be able to thank you enough for being mine, and for making my life so interesting, challenging, and hilariously fun. I lick you *and* love you. Derek, you're the steady heartbeat of it all—the eye of the hurricane. Thank you, thank you, thank you.

NOTES

Introduction

xii **thousands of fleeting micro-decisions:** Michael Yeomans et al., "A Practical Guide to Conversation Research," *Advances in Methods and Practices in Psychological Science* 6 (2023).

xiv **vastly complex and dynamic:** Michael Yeomans, Maurice Schweitzer, and Alison Wood Brooks, "The Conversational Circumplex: Identifying, Prioritizing, and Pursuing Informational and Relational Motives in Conversation," *Current Opinion in Psychology* 44 (2022): 293–302.

xiv **an ongoing act of co-creation:** Janet B. Bavelas, Linda Coates, and Trudy Johnson, "Listeners as Co-Narrators," *Journal of Personality and Social Psychology* 79, no. 6 (2000): 941.

xiv **can make a *big* difference:** Amit Kumar and Nicholas Epley, "Undersociality Is Unwise," *Journal of Consumer Psychology* 33, no. 1 (2023): 199–212.

xv **a dearth of feedback:** Robin M. Hogarth, Tomás Lejarraga, and Emre Soyer, "The Two Settings of Kind and Wicked Learning Environments," *Current Directions in Psychological Science* 24, no. 5 (2015): 379–85.

xv **we all transform into Narcissus during video calls:** Jeremy N. Bailenson, "Nonverbal Overload: A Theoretical Argument for the Causes of Zoom Fatigue" *Technology, Mind, and Behavior* 2, no. 1 (2021).

xvi **worse than they actually are:** Kenneth Savitsky, Nicholas Epley, and Thomas Gilovich, "Do Others Judge Us as Harshly as We Think? Overestimating the Impact of Our Failures, Shortcomings, and Mishaps," *Journal of Personality and Social Psychology* 81, no. 1 (2001): 44; Christopher Welker et al., "Pessimistic Assessments of Ability in Informal Conversation," *Journal of Applied Social Psychology* 53, no. 7 (2023): 555–69.

xviii **"difficult conversations" like negotiations:** Jared R. Curhan and Alex Pentland, "Thin Slices of Negotiation: Predicting Outcomes from Conversational Dynamics Within the First 5 Minutes," *Journal of Applied Psychology* 92, no. 3 (2007): 802.

xviii **speed dating:** Karen Huang et al., "It Doesn't Hurt to Ask: Question-Asking Increases Liking," *Journal of Personality and Social Psychology* 113, no. 3 (2017): 430.

xviii **parole hearings:** Grant E. Donnelly, Hanne Collins, and Alison Wood Brooks, "How Prisoner Apologies Influence Parole Decisions" (working).

xviii **negotiations:** Matteo Di Stasi, Alison Wood Brooks, and Jordi Quoidbach, "Asking Open-Ended Questions Increases Personal Gains in Negotiations" (working).

xviii **sales calls:** Alison Wood Brooks and Leslie K. John, "The Surprising Power of Questions," *Harvard Business Review* 96, no. 3 (2018): 60–67.

xviii **instant messaging, and face-to-face chinwags:** Alison Wood Brooks and Michael Yeomans, "Boomerasking: Answering Your Own Questions" (working); Michael Yeomans and Alison Wood Brooks, "Topic Preference Detection in Conversation: A Novel Approach to Understand Perspective Taking" (working).

xix **tone of the underlying relationship:** Yeomans, Schweitzer, and Brooks, "Conversational Circumplex."

xix **seeking the nearest exit:** Alison Wood Brooks and Maurice E. Schweitzer, "Can Nervous Nelly Negotiate? How Anxiety Causes Negotiators to Make Low First Offers, Exit Early, and Earn Less Profit," *Organizational Behavior and Human Decision Processes* 115, no. 1 (2011): 43–54.

xx **instructed to discuss nonwork topics:** Sean R. Martin et al., "Talking Shop: An Exploration of How Talking About Work Affects Our Initial Interactions," *Organizational Behavior and Human Decision Processes* 168 (2022).

xx **more likely to exit negotiations:** Brooks and Schweitzer, "Can Nervous Nelly Negotiate?"

xx **more likely to conceal information:** Gerben A. Van Kleef, Carsten K. W. De Dreu, and Antony S. R. Manstead, "The Interpersonal Effects of Anger and Happiness in Negotiations," *Journal of Personality and Social Psychology* 86, no. 1 (2004): 57.

xx **I designed a new course:** Marc Ethier, "The Most Interesting New MBA Courses at B-Schools This Year," *Poets & Quants*, September 22, 2019.

xxi **loneliness, one of today's greatest threats:** U.S. Department of Health and Human Services, "New Surgeon General Advisory Raises Alarm about the Devastating Impact of the Epidemic of Loneliness and Isolation in the United States" (press release), May 3, 2023, www.HHS.gov.

xxii **philosopher Paul Grice:** Paul Grice, *Studies in the Way of Words* (Cambridge, MA: Harvard University Press, 1991).

xxii **Levity helps us avoid boredom:** Erin C. Westgate and Timothy D. Wilson, "Boring Thoughts and Bored Minds: The MAC Model of Boredom and Cognitive Engagement," *Psychological Review* 125, no. 5 (2018): 689.

xxii **good conversation requires mutual attention:** Garriy Shteynberg, "A Collective Perspective: Shared Attention and the Mind," *Current Opinion in Psychology* 23 (2018): 93–97.

xxii **power of respect and good listening:** Hanne K. Collins, "When Listening Is Spoken," *Current Opinion in Psychology* 47 (2022).

xxiii **apologies may be the most powerful:** Daniel E. Forster et al., "Experimental Evidence That Apologies Promote Forgiveness by Communicating Relationship Value," *Scientific Reports* 11, no. 1 (2021).

xxiii **construct and maintain a shared reality:** Maya Rossignac-Milon et al., "Merged Minds: Generalized Shared Reality in Dyadic Relationships," *Journal of Personality and Social Psychology* 120, no. 4 (2021): 882.

Chapter 1: The Coordination Game

1 **exchange of words between two:** Michael Yeomans et al., "A Practical Guide to Conversation Research: How to Study What People Say to Each Other," *Advances in Methods and Practices in Psychological Science* 6, no. 4 (2023).

2 the Age of Conversation: Benedetta Craveri, *The Age of Conversation* (New York: New York Review Books, 2006).

2 "pleasurable" and "agreeable": Immanuel Kant, "§ 88. On the Highest Ethicophysical Good," in *Anthropology from a Pragmatic Point of View* (1798), trans. Victor Lyle Dowdell (Carbondale: Southern Illinois University Press, 1996).

2 "It is a certain manner of acting": Madame de Staël, *Germany, by Madame the Baroness de Staël-Holstein; with notes and appendices by O. W. Wright* (Boston: Houghton, Mifflin and Company, 1859); Barbara R. Hanning, "Conversation and Musical Style in the Late Eighteenth-Century Parisian Salon," *Eighteenth-Century Studies* 22, no. 4 (1989): 512–28; Margaret Bloom, "Conversation in the Writings of Mme de Staël," *PMLA* 48, no. 3 (1933): 861–66.

3 "republic of letters": Lorraine Daston, "The Ideal and Reality of the Republic of Letters in the Enlightenment," *Science in Context* 4, no. 2 (1991): 367–86.

3 "What Is Enlightenment?": Immanuel Kant, "An Answer to the Question: 'What Is Enlightenment?'" (1784).

4 "asked to table by the King in Königsberg": Manfred Kuehn, *Kant: A Biography* (Cambridge: Cambridge University Press, 2002), 357.

4 the rules of conversation he insisted on: Kant, "§ 88. On the Highest Ethicophysical Good"; Thomas de Quincey, "The 'Dinner Parties' of Immanuel Kant," in *The Last Days of Immanuel Kant* (1827), reprinted in *Anthologia* (2022), www.anthologialitt.com; Alix A. Cohen, "The Ultimate Kantian Experience: Kant on Dinner Parties," *History of Philosophy Quarterly* 25, no. 4 (2008): 315–36.

4 no know-it-alls: Cohen, "The Ultimate Kantian Experience."

5 "polite" or rarified: Jonathan Swift, "§ 17. Genteel Conversation, Directions to Servants, Argument Against Abolishing Christianity, and Other Pamphlets," in *From Steele and Addison to Pope and Swift*, vol. 9 of *The Cambridge History of English and American Literature in 18 Volumes*, ed. A. W. Ward and A. R. Waller (Cambridge University Press, 1907–21).

5 Dispatches from America: Stephen Miller, *Conversation: A History of a Declining Art* (New Haven, CT: Yale University Press, 2007), 203.

5 "the conduct of others": Adam Smith, *The Theory of Moral Sentiments and on the Origins of Languages*, ed. Dugald Stewart (London: Henry G. Bohn, 1853), 6.

6 call a *coordination game*: Russell Cooper, *Coordination Games* (Cambridge: Cambridge University Press, 1998); Thomas Schelling, *The Strategy of Conflict* (Cambridge, MA: Harvard University Press, 1960); Oskar Morgenstern and John von Neumann, *Theory of Games and Economic Behavior* (Princeton, NJ: Princeton University Press, 1944); and John F. Nash, "Equilibrium Points in N-Person Games," *Proceedings of the National Academy of Sciences* 36, no. 1 (1950): 48–49.

7 "focal point": Thomas Schelling, *The Strategy of Conflict* (Cambridge, MA: Harvard University Press, 1960).

8 relentless stream of coordination decisions: Yeomans et al., "Practical Guide."

8 if the other person wants: Thomas F. Pettigrew and Linda R. Tropp, "A Meta-Analytic Test of Intergroup Contact Theory," *Journal of Personality and Social Psychology* 90, no. 5 (2006): 751.

8 the question of what to talk about: Michael Yeomans and Alison Wood Brooks, "Topic Preference Detection in Conversation: A Novel Approach to Understand Perspective Taking" (working).

NOTES

8   **intimate conversations require:** Emma M. Templeton et al., "Long Gaps Between Turns Are Awkward for Strangers But Not for Friends," *Philosophical Transactions of the Royal Society B* 378, no. 1875 (2023).

9   **when should you leave:** Adam M. Mastroianni et al., "Do Conversations End When People Want Them To?," *Proceedings of the National Academy of Sciences* 118, no. 10 (2021): e2011809118; Elizabeth Stokoe, "The Sense of a Conversational Ending," Loughborough University, 2021, repository.lboro.ac.uk.

9   **what they say, and how they look:** Nicole Abi-Esber, Adam Mastroianni, and Alison Wood Brooks, "How Verbal, Nonverbal, and Paralinguistic Conversational Cues Inform Interpersonal Inference in Job Interviews" (working).

9   **"Now, then, gentlemen!":** De Quincey, "The 'Dinner Parties' of Immanuel Kant."

9   **"the jazz of human exchange":** Arlie Russell Hochschild, *Working in America* (London: Routledge, 2015), 29–36.

10   **"Jazz urges you to accept":** Wynton Marsalis and Geoffrey Ward, *Moving to Higher Ground: How Jazz Can Change Your Life* (New York: Random House, 2009).

10   **Emily Post:** Emily Post, *Etiquette in Society, in Business, in Politics, and at Home* (New York: Funk & Wagnalls, 1922).

10   **Dale Carnegie:** Dale Carnegie, *How to Win Friends and Influence People* (New York: Simon & Schuster, 1936).

11   **direct communication is neither viable:** Joy Hendry and Conrad William Watson, eds., *An Anthropology of Indirect Communication* (London: Routledge, 2001).

11   **"ordinary language":** John Langshaw Austin, *How to Do Things with Words* (Cambridge, MA: Harvard University Press, 1975).

12   **we always have at least one purpose:** Michael Yeomans, Maurice Schweitzer, and Alison Wood Brooks, "The Conversational Circumplex: Identifying, Prioritizing, and Pursuing Informational and Relational Motives in Conversation," *Current Opinion in Psychology* 44 (2022): 293–302.

12   **conversational compass:** Ibid.

15   **some of his priorities conflict:** Gráinne M. Fitzsimons, Eli J. Finkel, and Michelle R. vanDellen, "Transactive Goal Dynamics," *Psychological Review* 122, no. 4 (2015): 648; Henri Barki and Jon Hartwick, "Conceptualizing the Construct of Interpersonal Conflict," *International Journal of Conflict Management* 15, no. 3 (2004): 216–44.

16   **improvisational, constantly changing nature:** Yeomans et al., "Practical Guide."

17   **key elements of his theory:** Paul Grice, *Studies in the Way of Words* (Cambridge, MA: Harvard University Press, 1991).

18   **we violate Grice's elegant maxims:** Suellen Rundquist, "Indirectness: A Gender Study of Flouting Grice's Maxims," *Journal of Pragmatics* 18, no. 5 (1992): 431–49.

18   **we prioritize kindness over honesty:** Emma E. Levine and Matthew J. Lupoli, "Prosocial Lies: Causes and Consequences," *Current Opinion in Psychology* 43 (2022): 335–40; Emma E. Levine and Maurice E. Schweitzer, "Prosocial Lies: When Deception Breeds Trust," *Organizational Behavior and Human Decision Processes* 126 (2015): 88–106.

18   **brief answers might arouse suspicion:** Leslie K. John, Kate Barasz, and Michael I. Norton, "Hiding Personal Information Reveals the Worst," *Proceedings of the National Academy of Sciences* 113, no. 4 (2016): 954–59.

18	**We shouldn't always be *relevant*:** Yeomans and Brooks, "Topic Preference Detection in Conversation."
18	**filler words are important:** Herbert H. Clark and Jean E. Fox Tree, "Using Uh and Um in Spontaneous Speaking," *Cognition* 84, no. 1 (2002): 73–111.
19	**a gifted eavesdropper:** Francesco Ranci, "The Unfinished Business of Erving Goffman: From Marginalization Up Towards the Elusive Center of American Sociology," *American Sociologist* 52, no. 2 (2021): 390–419; Philip Manning, *Erving Goffman and Modern Sociology* (Hoboken, NJ: John Wiley & Sons, 2013).
19	**small-scale interactions of everyday life:** Erving Goffman, *The Presentation of Self in Everyday life* (University of Edinburgh Social Sciences Research Centre, 1959); Erving Goffman, *Strategic Interaction* (Philadelphia: University of Pennsylvania Press, 1969); Erving Goffman, *Forms of Talk* (Philadelphia: University of Pennsylvania Press, 1981).
19	**Conversation Analysis:** Emanuel A. Schegloff, *Sequence Organization in Interaction: A Primer in Conversation Analysis* (Cambridge: Cambridge University Press, 2007), vol. 1.
19	**one recent study, sniffing:** Elliott M Hoey, "Waiting to Inhale: On Sniffing in Conversation," *Research on Language and Social Interaction* 53, no. 1 (2020): 118–39.
19	**need to know the participants' purposes:** Yeomans, Schweitzer, and Brooks, "Conversational Circumplex"; Stokoe, "Sense of a Conversational Ending."
19	**Bigger claims required bigger data:** Yeomans et al., "Practical Guide."
20	**treat words like numbers:** Ibid.; Aravind K. Joshi, "Natural Language Processing," *Science* 253, no. 5025 (1991): 1242–49.
20	**about sixteen thousand words per day:** Matthias R. Mehl et al., "Are Women Really More Talkative Than Men?," *Science* 317, no. 5834 (2007): 82.
20	**psychologist Gillian Sandstrom:** Alecia J. Carter et al., "Women's Visibility in Academic Seminars: Women Ask Fewer Questions Than Men," *PLOS One* 13, no. 9 (2018): e0202743.
21	**over two million years:** Marc D. Hauser et al., "The Mystery of Language Evolution," *Frontiers in Psychology* 5 (2014): 401.
21	**since we were toddlers:** Jenny R. Saffran, Ann Senghas, and John C. Trueswell, "The Acquisition of Language by Children," *Proceedings of the National Academy of Sciences* 98, no. 23 (2001): 12874–75.
21	**not only in my research:** Karen Huang et al., "It Doesn't Hurt to Ask: Question-Asking Increases Liking," *Journal of Personality and Social Psychology*, no. 3 (2017): 430; T. Bradford Bitterly and Maurice E. Schweitzer, "The Impression Management Benefits of Humorous Self-Disclosures: How Humor Influences Perceptions of Veracity," *Organizational Behavior and Human Decision Processes* 151 (2019): 73–89; Matteo Di Stasi, Alison Wood Brooks, and Jordi Quoidbach, "Asking Open-Ended Questions Increases Personal Gains in Negotiations" (working); Alison Wood Brooks and Michael Yeomans, "Boomerasking: Answering Your Own Questions" (working); Yeomans and Brooks, "Topic Preference Detection in Conversation"; Nicole Abi-Esber et al., "The Power of Preparation: Brainstorming Flexible Topics Before Conversations Begin" (working).

Chapter 2: T Is for Topics

28 **"Mine was no ordinary Greek"**: Primo Levi, *The Reawakening,* trans. Stuart Woolf (London: Bodley Head, 1965), 45.

28 **"loose jointed mind"**: John Sutherland, *Stephen Spender: The Authorized Biography* (New York: Viking, 2004), 251.

29 ***verbal*** **conversational content**: Michael Yeomans et al., "A Practical Guide to Conversation Research," *Advances in Methods and Practices in Psychological Science* 6 (2023).

29 **it's the topics that we**: Michael Yeomans and Alison Wood Brooks, "Topic Preference Detection in Conversation: A Novel Approach to Understand Perspective Taking" (working).

30 **we asked a thousand people**: Ibid.

31 **develop topics based on your relationship**: Michael Yeomans, Maurice Schweitzer, and Alison Wood Brooks, "The Conversational Circumplex: Identifying, Prioritizing, and Pursuing Informational and Relational Motives in Conversation," *Current Opinion in Psychology* 44 (2022): 293–302.

32 **Topic management is why**: Nicole Abi-Esber et al., "The Power of Preparation: Brainstorming Flexible Topics Before Conversations Begin" (working).

32 **27 percent of people report spending**: Ibid.

33 **"system 1 thinking"**: Daniel Kahneman, *Thinking, Fast and Slow* (New York: Macmillan, 2011).

33 **prove how great we are**: Mark R. Leary and Robin M. Kowalski, "Impression Management: A Literature Review and Two-Component Model," *Psychological Bulletin* 107, no. 1 (1990): 34.

33 **or how right we are**: Julia A. Minson, Frances S. Chen, and Catherine H. Tinsley, "Why Won't You Listen to Me? Measuring Receptiveness to Opposing Views," *Management Science* 66, no. 7 (2020): 3069–94.

33 **(our natural egocentrism)**: Nicholas Epley and Eugene M. Caruso, "Egocentric Ethics," *Social Justice Research* 17 (2004): 171–87; Nicholas Epley et al., "Perspective Taking as Egocentric Anchoring and Adjustment," *Journal of Personality and Social Psychology* 87, no. 3 (2004): 327.

34 **"cognitive offloading"**: Evan F. Risko and Sam J. Gilbert, "Cognitive Offloading," *Trends in Cognitive Sciences* 20, no. 9 (2016): 676–88.

34 **trying to listen and respond**: Hanne K. Collins et al., "Conveying and Detecting Listening During Live Conversation," *Journal of Experimental Psychology: General* 153, no. 2 (2024): 473–94.

35 **usually improves immensely**: Abi-Esber et al., "Power of Preparation."

35 **improves networking success**: Mindy Truong, Nathanael J. Fast, and Jennifer Kim, "It's Not What You Say, It's How You Say It: Conversational Flow As a Predictor of Networking Success," *Organizational Behavior and Human Decision Processes* 158 (2020): 1–10.

35 **cover *more* topics**: Gus Cooney et al., "Switching Topics More Frequently Makes Boring Conversations Better" (working).

36 **learning more about your partner's perspective**: Tal Eyal, Mary Steffel, and Nicholas Epley, "Perspective Mistaking: Accurately Understanding the Mind of Another Requires Getting Perspective, Not Taking Perspective," *Journal of Personality and Social Psychology* 114, no. 4 (2018): 547; Karen Huang et al., "It Doesn't Hurt to

Ask: Question-Asking Increases Liking," *Journal of Personality and Social Psychology* 113, no. 3 (2017): 430; Matteo Di Stasi, Alison Wood Brooks, and Jordi Quoidbach, "Asking Open-Ended Questions Increases Personal Gains in Negotiations" (working).

38 **the particular torture:** Matthias R. Mehl et al., "Eavesdropping on Happiness: Well-Being Is Related to Having Less Small Talk and More Substantive Conversations," *Psychological Science* 21, no. 4 (2010): 539–41.

38 **good reasons to feel this way:** Ibid.

38 **small talk doom spiral:** Michael Kardas, Amit Kumar, and Nicholas Epley, "Overly Shallow?: Miscalibrated Expectations Create a Barrier to Deeper Conversation," *Journal of Personality and Social Psychology* 122, no. 3 (2022): 367.

38 **initial proving ground:** Justine Coupland, "Small Talk: Social Functions," *Research on Language and Social Interaction* 36, no. 1 (2003): 1–6.

39 **topic forethought can pay dividends:** Abi-Esber et al., "Power of Preparation."

40 **"House on Loon Lake":** Ira Glass, "House on Loon Lake," *This American Life,* November 16, 2001.

42 **stay on the boat too long:** Kardas, Kumar, and Epley, "Overly Shallow?"; Yeomans and Brooks, "Topic Preference Detection in Conversation."

42 **"The merest everyday speech-morsel":** James Parker, "An Ode to Small Talk," *Atlantic,* October 2020.

43 **One goal is to move up:** Kardas, Kumar, and Epley, "Overly Shallow?"

44 **don't get stuck for too long:** Cooney et al., "Switching Topics More Frequently."

44 **personal experience, expertise:** Diana I. Tamir and Jason P. Mitchell, "Disclosing Information About the Self Is Intrinsically Rewarding," *Proceedings of the National Academy of Sciences* 109, no. 21 (2012): 8038–43.

44 **aiming to make topics concrete:** Michael Yeomans, "A Concrete Example of Construct Construction in Natural Language," *Organizational Behavior and Human Decision Processes* 162 (2021): 81–94.

44 **personal disclosure tends to trigger:** Susan Sprecher et al., "Taking Turns: Reciprocal Self-Disclosure Promotes Liking in Initial Interactions," *Journal of Experimental Social Psychology* 49, no. 5 (2013): 860–66; Lynn C. Miller and David A. Kenny, "Reciprocity of Self-Disclosure at the Individual and Dyadic Levels: A Social Relations Analysis," *Journal of Personality and Social Psychology* 50, no. 4 (1986): 713.

45 **motor of vulnerability and trust:** Ann-Marie Nienaber, Marcel Hofeditz, and Philipp Daniel Romeike, "Vulnerability and Trust in Leader-Follower Relationships," *Personnel Review* 44, no. 4 (2015): 567–91.

45 **uniquely suited to a particular combination:** Maya Rossignac-Milon et al., "Merged Minds: Generalized Shared Reality in Dyadic Relationships," *Journal of Personality and Social Psychology* 120, no. 4 (2021): 882.

45 **secret your partner wants to share:** Michael L. Slepian and Katharine H. Greenaway, "The Benefits and Burdens of Keeping Others' Secrets," *Journal of Experimental Social Psychology* 78 (2018): 220–32; Michael L. Slepian and Edythe Moulton-Tetlock, "Confiding Secrets and Well-Being," *Social Psychological and Personality Science* 10, no. 4 (2019): 472–84; Michael L. Slepian and James N. Kirby, "To Whom Do We Confide Our Secrets?," *Personality and Social Psychology Bulletin* 44, no. 7 (2018): 1008–23; Leslie John, Michael L. Slepian, and Diana Tamir, "Tales of Two Motives: Disclosure and Concealment," *Current Opinion in Psychology* 31 (2020).

45 **specific problem or triumph:** Alison Wood Brooks et al., "Mitigating Malicious Envy: Why Successful Individuals Should Reveal Their Failures," *Journal of Experimental Psychology: General* 148, no. 4 (2019): 667; Annabelle R. Roberts, Emma E. Levine, and Ovul Sezer, "Hiding Success," *Journal of Personality and Social Psychology* 120, no. 5 (2021): 1261.

45 **focus on getting to know each other:** Emma M. Templeton et al., "Long Gaps Between Turns Are Awkward for Strangers But Not for Friends," *Philosophical Transactions of the Royal Society B* 378, no. 1875 (2023); Stav Atir, Kristina A. Wald, and Nicholas Epley, "Talking with Strangers Is Surprisingly Informative," *Proceedings of the National Academy of Sciences* 119, no. 34 (2022): e2206992119.

46 **Forethought—thinking ahead:** Abi-Esber et al., "Power of Preparation."

46 **I can call back to?:** Hanne K. Collins and Alison Wood Brooks, "Call-backs in Conversation" (working).

48 **the *when* is just as important:** Grant Packard, Yang Li, and Jonah Berger, "When Language Matters," *Journal of Consumer Research* (2023): ucad080.

48 **interest and engagement flag:** Erin C. Westgate and Timothy D. Wilson, "Boring Thoughts and Bored Minds: The MAC Model of Boredom and Cognitive Engagement," *Psychological Review* 125, no. 5 (2018): 689.

48 **(the birth of the uncomfortable silence):** Dennis Kurzon, "Towards a Typology of Silence," *Journal of Pragmatics* 39, no. 10 (2007): 1673–88; Emma M. Templeton et al., "Fast Response Times Signal Social Connection in Conversation," *Proceedings of the National Academy of Sciences* 119, no. 4 (2022): e2116915119.

49 **a study using pairs of strangers:** Cooney et al., "Switching Topics More Frequently."

50 **Does a broad range come at the expense:** Yeomans and Brooks, "Topic Preference Detection in Conversation."

50 **people read their partners' cues:** Ibid.

51 **"co-narrated":** Janet B. Bavelas, Linda Coates, and Trudy Johnson, "Listeners as Co-Narrators," *Journal of Personality and Social Psychology* 79, no. 6 (2000): 941; Hanne K. Collins, "When Listening Is Spoken," *Current Opinion in Psychology* 47 (2022).

53 **"create an atmosphere":** Levi, *Reawakening*, 45.

Chapter 3: A Is for Asking

56 **Carrie Fisher was two years old:** Dave Itzkoff, "Carrie Fisher, Child of Hollywood and 'Star Wars' Royalty, Dies at 60," *New York Times,* December 27, 2016; Carrie Fisher and Rob Delaney, *Carrie Fisher: The Memoirs* (New York: Simon & Schuster, 2018).

56 **a novel, *Postcards from the Edge:*** Carrie Fisher, *Postcards from the Edge* (New York; Simon & Schuster, 2008).

57 **a successful screenplay:** Carrie Fisher, *Postcards from the Edge* screenplay (1988).

57 **semiautobiographical novel, *Surrender the Pink:*** Carrie Fisher, *Surrender the Pink* (New York: Simon & Schuster, 1990).

57 **a guest on the program *Fresh Air:*** "Actress and Author Carrie Fisher," *Fresh Air,* NPR, February 21, 1997.

58 **The next one came in 2004:** "Actress and Novelist Carrie Fisher," *Fresh Air,* NPR, February 4, 2004.

58 **the third in 2016:** "Carrie Fisher Opens Up About 'Star Wars,' The Gold Bikini and Her On-Set Affair," *Fresh Air,* NPR, Novemer 28, 2016.

61 **Thanks to her meticulous preparation:** Terry Gross, *All I Did Was Ask: Conversations with Writers, Actors, Musicians, and Artists* (New York: Hachette, 2004).

61 **Questions are the most powerful tool:** Matteo Di Stasi, Alison Wood Brooks, and Jordi Quoidbach, "Asking Open-Ended Questions Increases Personal Gains in Negotiations" (working); Karen Huang et al., "It Doesn't Hurt to Ask: Question-Asking Increases Liking," *Journal of Personality and Social Psychology* 113, no. 3 (2017): 430.

61 **parallel monologues between speakers:** "Behavior: The Art of Not Listening," *Time,* January 24, 1969; Janet B. Bavelas, Linda Coates, and Trudy Johnson, "Listeners as Co-narrators," *Journal of Personality and Social Psychology* 79.6 (2000): 941.

61 **Asking questions sets your partner up:** Julia A. Minson et al., "Eliciting the Truth, the Whole Truth, and Nothing But the Truth: The Effect of Question Phrasing on Deception," *Organizational Behavior and Human Decision Processes* 147 (2018): 76–93; Alison Wood Brooks and Leslie K. John, "The Surprising Power of Questions," *Harvard Business Review* 96, no. 3 (2018): 60–67.

61 **Questions are the best-known way to learn:** Tal Eyal, Mary Steffel, and Nicholas Epley, "Perspective Mistaking: Accurately Understanding the Mind of Another Requires Getting Perspective, Not Taking Perspective," *Journal of Personality and Social Psychology* 114, no. 4 (2018): 547.

61 **singular evolutionary gift:** Eric Hedin, "Asking Questions and Human Exceptionalism," *Evolution News,* August 7, 2023.

61 **bonobos, have learned to communicate:** Lindsay Stern, "What Can Bonobos Teach Us about the Nature of Language?," *Smithsonian Magzine,* July 2020.

62 **correlates with all kinds of positive outcomes:** Huang et al., "It Doesn't Hurt to Ask"; Michael Yeomans et al., "It Helps to Ask: The Cumulative Benefits of Asking Follow-Up Questions," *Journal of Personality and Social Psychology* 117, no. 6 (2019): 1139–44; Di Stasi, Brooks, and Quoidbach, "Asking Open-Ended Questions"; Brooks and John, "Surprising Power of Questions"; Grant E. Donnelly, Hanne Collins, and Alison Wood Brooks, "How Prisoner Apologies Influence Parole Decisions" (working).

62 **question askers learn more information:** Stav Atir, Kristina A. Wald, and Nicholas Epley, "Talking with Strangers Is Surprisingly Informative," *Proceedings of the National Academy of Sciences* 119, no. 34 (2022): e2206992119.

62 **Asking more questions increases information exchange:** Di Stasi, Brooks, and Quoidbach, "Asking Open-Ended Questions."

62 **People who ask more questions:** Huang et al., "It Doesn't Hurt to Ask."

63 **people don't do nearly enough of it:** Ibid.

64 **vastly *overestimate* how many questions:** Di Stasi, Brooks, and Quoidbach, "Asking Open-Ended Questions"; Huang et al., "It Doesn't Hurt to Ask."

64 **people who had asked lots of questions:** Huang et al., "It Doesn't Hurt to Ask."

64 **They're not getting clear feedback:** Robin M. Hogarth, Tomás Lejarraga, and

Emre Soyer, "The Two Settings of Kind and Wicked Learning Environments," *Current Directions in Psychological Science* 24, no. 5 (2015): 379–85.

65 **we just need to *ask more* questions:** Huang et al., "It Doesn't Hurt to Ask."

65 **predictions about how their questions would be:** Einav Hart, Eric M. VanEpps, and Maurice E. Schweitzer, "The (Better Than Expected) Consequences of Asking Sensitive Questions," *Organizational Behavior and Human Decision Processes* 162 (2021): 136–54.

68 **Most of us are afraid of asking:** Ibid.

69 **And getting personal makes us vulnerable:** Tami Kim, Kate Barasz, and Leslie K. John, "Consumer Disclosure," *Consumer Psychology Review* 4, no. 1 (2021): 59–69; Li Jiang et al., "Fostering Perceptions of Authenticity via Sensitive Self-Disclosure," *Journal of Experimental Psychology: Applied* 28, no. 4 (2022): 898.

69 **opens us to trust, enjoyment, and love:** Susan Sprecher, "Closeness and Other Affiliative Outcomes Generated from the Fast Friends Procedure: A Comparison with a Small-Talk Task and Unstructured Self-Disclosure and the Moderating Role of Mode of Communication," *Journal of Social and Personal Relationships* 38, no. 5 (2021): 1452–71; Arthur Aron et al., "The Experimental Generation of Interpersonal Closeness: A Procedure and Some Preliminary Findings," *Personality and Social Psychology Bulletin* 23, no. 4 (1997): 363–77.

69 **several other exercises:** Deepak Malhotra and Max Bazerman, *Negotiation Genius: How to Overcome Obstacles and Achieve Brilliant Results at the Bargaining Table and Beyond* (New York: Bantam Books, 2007).

70 **invite information from a partner:** Di Stasi, Brooks, and Quoidbach, "Asking Open-Ended Questions."

70 **revealing the motives of the asker:** Minson et al., "Eliciting the Truth, the Whole Truth."

70 **steering us toward the next turn:** Michael Yeomans and Alison Wood Brooks, "Topic Preference Detection in Conversation: A Novel Approach to Understand Perspective Taking" (working).

71 **four dominant question types emerged:** Huang et al.,"It Doesn't Hurt to Ask."

72 **asking more questions increases likability:** Ibid.

76 **If you had to do it all over again:** Monica Lewinsky, interview by Barbara Walters, March 3, 1999, https://www.youtube.com/watch?v=vUUATD_pfYE.

78 **our study of speed daters:** Huang et al.,"It Doesn't Hurt to Ask."

80 **more questions are good:** Brooks and John, "Surprising Power of Questions"; Chris Orlob, "4 Tips for Nailing Your Sales Discovery Calls," *Gong,* June 18, 2017.

81 **"How was your weekend?":** Alison Wood Brooks and Michael Yeomans, "Boomerasking: Answering Your Own Questions" (working).

82 **our deepest, darkest bad intentions:** Minson et al., "Eliciting the Truth, the Whole Truth."

83 **perception of positive or negative intent:** Daniel L. Ames and Susan T. Fiske, "Perceived Intent Motivates People to Magnify Observed Harms," *Proceedings of the National Academy of Sciences* 112, no. 12 (2015): 3599–605.

84 **people respond to open-ended questions:** Di Stasi, Brooks, and Quoidbach, "Asking Open-Ended Questions."

85 **Anita Pomerantz has called "candidate answers":** Anita Pomerantz, "Offering a

Candidate Answer: An Information Seeking Strategy," *Communications Monographs* 55, no. 4 (1988): 360–73.

85 **open-ended and closed questions during negotiations:** Di Stasi, Brooks, and Quoidbach, "Asking Open-Ended Questions."

86 **it's a good way to get there:** Aron et al., "Experimental Generation of Interpersonal Closeness."

Chapter 4: L Is for Levity

88 **romantic dates, interpersonal attraction:** Eli J. Finkel, *The All-or-Nothing Marriage: How the Best Marriages Work* (New York: Penguin, 2019); Paul W. Eastwick, Eli J. Finkel, and Samantha Joel, "Mate Evaluation Theory," *Psychological Review* 130, no. 1 (2023): 211.

88 **recording and analyzing the transcripts:** Rajesh Ranganath, Dan Jurafsky, and Daniel A. McFarland, "Detecting Friendly, Flirtatious, Awkward, and Assertive Speech in Speed-Dates," *Computer Speech and Language* 27, no. 1 (2013): 89–115.

89 **"Hollaback Girl" by Gwen Stefani:** "Billboard Hot 100TM: Week of June 11, 2005," *Billboard*.

89 **Brad Pitt and Jennifer Aniston divorced:** Leslie Bennetts, "The Unsinkable Jennifer Aniston: Vanity Fair," *Vanity Fair,* September 2005.

92 **There's a risk to overanalyzing:** Nicole Abi-Esber, Adam Mastroianni, and Alison Wood Brooks, "How Verbal, Nonverbal, and Paralinguistic Conversational Cues Shape Impressions" (working).

92 **clues to their chemistry:** Harry T. Reis, Annie Regan, and Sonja Lyubomirsky, "Interpersonal Chemistry: What Is It, How Does It Emerge, and How Does It Operate?," *Perspectives on Psychological Science* 17, no. 2 (2022): 530–58.

93 **The prospect of an awkward silence:** Emma M. Templeton et al., "Long Gaps Between Turns Are Awkward for Strangers But Not for Friends," *Philosophical Transactions of the Royal Society B* 378, no. 1875 (2023).

94 **complex constellation of human emotions:** James A. Russell, "A Circumplex Model of Affect," *Journal of Personality and Social Psychology* 39, no. 6 (1980): 1161.

94 **Our minds wander:** Jonathan Smallwood and Jonathan W. Schooler, "The Science of Mind Wandering: Empirically Navigating the Stream of Consciousness," *Annual Review of Psychology* 66 (2015): 487–518.

94 **We make less eye contact:** Daniel Smilek, Jonathan S. A. Carriere, and J. Allan Cheyne, "Out of Mind, Out of Sight: Eye Blinking as Indicator and Embodiment of Mind Wandering," *Psychological Science* 21, no. 6 (2010): 786–89.

94 **Our words get less positive:** Grant Packard, Yang Li, and Jonah Berger, "When Language Matters," *Journal of Consumer Research,* December 22, 2023; Jonah Berger, Wendy W. Moe, and David A. Schweidel, "What Holds Attention? Linguistic Drivers of Engagement," *Journal of Marketing* 87, no. 5 (2023): 793–809.

94 **we interrupt less:** Hanne K. Collins et al., "Conveying and Detecting Listening During Live Conversation," *Journal of Experimental Psychology: General* 153, no. 2 (2023): 473–94.

95 **We lean forward:** Jinni A. Harrigan, Thomas E. Oxman, and Robert Rosenthal, "Rapport Expressed Through Nonverbal Behavior," *Journal of Nonverbal Behavior* 9

(1985): 95–110; Sophie Wohltjen and Thalia Wheatley, "Eye Contact Marks the Rise and Fall of Shared Attention in Conversation," *Proceedings of the National Academy of Sciences* 118, no. 37 (2021): e2106645118.

95 **reactions of back-channel feedback:** Herbert H. Clark and Susan E. Brennan, "Grounding in Communication," in L. B. Resnick, J. M. Levine, and S. D. Teasley, eds., *Perspectives on Socially Shared Cognition*, 127–49 (New York: American Psychological Association, 1991).

95 **We respond more quickly:** Emma M. Templeton et al., "Fast Response Times Signal Social Connection in Conversation," *Proceedings of the National Academy of Sciences* 119, no. 4 (2022): e2116915119.

95 **We even interrupt more often:** Collins et al., "Conveying and Detecting Listening."

96 **randomly selected people feel happier:** Andrew J. Oswald, Eugenio Proto, and Daniel Sgroi, "Happiness and Productivity," *Journal of Labor Economics* 33, no. 4 (2015): 789–822.

96 **feeling happy provokes greater creativity:** Teresa M. Amabile et al., "Affect and Creativity at Work," *Administrative Science Quarterly* 50, no. 3 (2005): 367–403; Teresa Amabile and Steven Kramer, *The Progress Principle: Using Small Wins to Ignite Joy, Engagement, and Creativity at Work* (Cambridge, MA: Harvard Business Press, 2011).

96 **generate more ideas—with a higher proportion:** Alison Wood Brooks, "Get Excited: Reappraising Pre-Performance Anxiety as Excitement," *Journal of Experimental Psychology: General* 143, no. 3 (2014): 1144.

96 **"There is no better trigger":** Walter Benjamin, "The Author as Producer," in Victor Burgin, ed., *Thinking Photography* (Red Globe Press, 1982): 15–31.

96 **When we're happy and engaged:** Barbara L. Fredrickson, "The Role of Positive Emotions in Positive Psychology: The Broaden-and-Build Theory of Positive Emotions," *American Psychologist* 56, no. 3 (2001): 218.

96 **One good laugh can relieve physical tension:** Rachel Hajar, "Laughter in Medicine," *Heart Views* 24, no. 2 (2023): 124.

96 **Our recovery from stress:** Anthony D. Ong, "Pathways Linking Positive Emotion and Health in Later Life," *Current Directions in Psychological Science* 19, no. 6 (2010): 358–62.

97 **Psychological safety—the belief:** Amy C. Edmondson, *The Fearless Organization: Creating Psychological Safety in the Workplace for Learning, Innovation, and Growth* (Hoboken, NJ: John Wiley & Sons, 2018).

97 **These cues of levity:** Tami Kim, Kate Barasz, and Leslie K. John, "Consumer Disclosure," *Consumer Psychology Review* 4, no. 1 (2021): 59–69.

98 **most of us forget or hesitate:** Jennifer Aaker and Naomi Bagdonas, *Humor, Seriously: Why Humor Is a Secret Weapon in Business and Life (And How Anyone Can Harness It. Even You)* (New York: Crown Currency, 2021); Brad Bitterly and Alison Wood Brooks, "Sarcasm, Self-Deprecation, and Inside Jokes: A User's Guide to Humor at Work," *Harvard Business Review*, July–August 2020.

98 **frivolous, incompetent, or disrespectful:** Bradford T. Bitterly, Alison Wood Brooks, and Maurice E. Schweitzer, "Risky Business: When Humor Increases and Decreases Status," *Journal of Personality and Social Psychology* 112, no. 3 (2017): 431.

99 **"broaden and build" theory:** Fredrickson, "Role of Positive Emotions."

99 **"Yes, and" mantra:** Kelly Leonard and Tom Yorton, *Yes, and: How Improvisation*

Reverses "No, But" Thinking and Improves Creativity and Collaboration (New York: HarperCollins, 2015).

100 **anxiety about conversational humor:** Bitterly, Brooks, and Schweitzer, "Risky Business."

101 **Benign Violation Theory of Humor:** A. Peter McGraw and Caleb Warren, "Benign Violations: Making Immoral Behavior Funny," *Psychological Science* 21, no. 8 (2010): 1141–49.

102 **how we position ourselves:** Aaker and Bagdonas, *Humor, Seriously*.

103 **"I feel like Elizabeth Taylor's":** Yitzi Weiner, "Social Impact Authors: How & Why Author Ric Keller Is Helping to Change Our World," *Medium*, August 19, 2022.

103 **make fun of Frank Sinatra to his face:** "Don Rickles Roasts Frank Sinatra," https://www.youtube.com/watch?v=K-KeTNU-ods.

104 **less costly to flop for being too benign:** Bitterly, Brooks, and Schweitzer, "Risky Business."

105 **"compare and contrast":** Aaker and Bagdonas *Humor, Seriously*.

107 **views of the joker's confidence rose:** Bitterly, Brooks, and Schweitzer, "Risky Business."

107 **"bearded, congenial, and wisecracking presence":** Benedict Carey, "Robert Provine, an Authority on Laughter, Is Dead at 76," *New York Times*, October 28, 2019.

107 **"What would the visitor make":** Robert R. Provine, *Laughter: A Scientific Investigation* (New York: Viking, 2001).

109 **"situation modification":** James J. Gross, "Emotion Regulation: Current Status and Future Prospects," *Psychological Inquiry* 26, no. 1 (2015): 1–26.

110 ***call-backs*—explicit references:** Hanne K. Collins and Alison Wood Brooks, "Callbacks in Conversation" (working).

110 **the joy of rediscovery:** Ting Zhang et al., "A 'Present' for the Future: The Unexpected Value of Rediscovery," *Psychological Science* 25, no. 10 (2014): 1851–60.

112 **Off-the-wall topics:** Nicole Abi-Esber et al., "The Power of Preparation: Brainstorming Flexible Topics Before Conversations Begin" (working); Gus Cooney et al., "Switching Topics More Frequently Makes Boring Conversations Better" (working); Michael Yeomans and Alison Wood Brooks, "Topic Preference Detection in Conversation: A Novel Approach to Understand Perspective Taking" (working).

112 **Hearing laughter from a partner:** Matthew Gervais and David Sloan Wilson, "The Evolution and Functions of Laughter and Humor: A Synthetic Approach," *Quarterly Review of Biology* 80, no. 4 (2005): 395–430; Robert R. Provine, "Contagious Laughter: Laughter Is a Sufficient Stimulus for Laughs and Smiles," *Bulletin of the Psychonomic Society* 30, no. 1 (1992): 1–4.

113 **"Voiced, songlike laughs":** Jo-Anne Bachorowski, Moria J. Smoski, and Michael J. Owren, "The Acoustic Features of Human Laughter," *Journal of the Acoustical Society of America* 110, no. 3 (2001): 1581–97; Jo-Anne Bachorowski and Michael J. Owren, "Not All Laughs Are Alike: Voiced but Not Unvoiced Laughter Readily Elicits Positive Affect," *Psychological Science* 12, no. 3 (2001): 252–257.

113 **people we think of as "funny":** Robert R. Provine, "Laughter," *American Scientist* 84, no. 1 (1996): 38–45; Robert R. Provine, "Laughing, Tickling, and the Evolution of Speech and Self," *Current Directions in Psychological Science* 13, no. 6 (2004): 215–18.

113 **over 70 percent of laughter occurrences:** T. Bradford Bitterly, Alison Wood Brooks, Maurice Schweitzer, and Jennifer Aaker, "Gender and Laughter" (working).

114 **"We are contacting you directly"**: Elaine Chan and Jaideep Sengupta, "Insincere Flattery Actually Works: A Dual Attitudes Perspective," *Journal of Marketing Research* 47, no. 1 (2010): 122–33.

115 **People think that giving compliments**: Erica J. Boothby, and Vanessa K. Bohns, "Why a Simple Act of Kindness Is Not as Simple as It Seems: Underestimating the Positive Impact of Our Compliments on Others," *Personality and Social Psychology Bulletin* 47, no. 5 (2021): 826–40; Vanessa Bohns, *You Have More Influence Than You Think: How We Underestimate Our Powers of Persuasion, and Why It Matters* (WW Norton, 2021).

116 **match their ideal partner preferences**: Lorne Campbell and Garth J. O. Fletcher, "Romantic Relationships, Ideal Standards, and Mate Selection," *Current Opinion in Psychology* 1 (2015): 97–100.

117 **asked people on their deathbeds**: Aaker and Bagdonas, *Humor, Seriously*.

Chapter 5: K Is for Kindness

118 **Gloria Vanderbilt, an heiress**: Robert D. McFadden, "Gloria Vanderbilt Dies at 95; Built a Fashion Empire," *New York Times*, June 17, 2019.

119 **The conversation between Cooper and Colbert**: "Stephen Colbert and Anderson Cooper's Beautiful Conversation About Grief," YouTube (2019), https://www.youtube.com/watch?v=YB46h1koicQ.

121 **"best TV moment"**: Hank Stuever, "2019's Best TV Moment? It Was Stephen Colbert Answering Anderson Cooper's Question About Grief," *Washington Post*, December 23, 2019.

121 **"This is something we all go through"**: Robin Pogrebin, "Anderson Cooper Explores Grief and Loss in Deeply Personal Podcast," *New York Times*, November 28, 2022.

123 **state of "extreme egocentrism"**: David Elkind, "Egocentrism in Adolescence," *Child Development* 38, no. 4 (1967): 1025–34.

124 **sabotages our ability to choose good topics**: Michael Yeomans and Alison Wood Brooks, "Topic Preference Detection in Conversation: A Novel Approach to Understand Perspective Taking" (working).

124 **egocentric we are during conversation**: Becky Ka Ying Lau et al., "The Extreme Illusion of Understanding," *Journal of Experimental Psychology: General* 151, no. 11 (2022): 2957.

125 **"The single biggest problem with communication"**: Conor Kenny, "The Single Biggest Problem in Communication Is the Illusion That It Has Taken Place," *Irish Times*, November 9, 2020.

126 **We all have our own needs**: Michael Yeomans, Maurice Schweitzer, and Alison Wood Brooks, "The Conversational Circumplex: Identifying, Prioritizing, and Pursuing Informational and Relational Motives in Conversation," *Current Opinion in Psychology* 44 (2022): 293–302.

127 **written by Otis Redding**: Jacob Uitti, "Who Wrote the Historic Song 'Respect'?" AmericanSongwriter.com, 2023.

127 **camera footage from 981 traffic stops**: Rob Voigt et al., "Language from Police Body Camera Footage Shows Racial Disparities in Officer Respect," *Proceedings of the National Academy of Sciences* 114, no. 25 (2017): 6521–26.

NOTES

128 **twenty-two categories:** Laurence R. Horn and Gregory L. Ward, eds., *The Handbook of Pragmatics* (Oxford: Blackwell, 2004).

130 **"The main hope of harmony":** Amartya Sen, *Identity and Violence: The Illusion of Destiny* (New York: W.W. Norton, 2007).

133 **amygdala hijacking:** Daniel Goleman, *Emotional Intelligence*, 10th ed. (New York: Bantam Books, 2007).

133 **"like the common cold":** Trevor Foulk, Andrew Woolum, and Amir Erez, "Catching Rudeness Is Like Catching a Cold: The Contagion Effects of Low-Intensity Negative Behaviors," *Journal of Applied Psychology* 101, no. 1 (2016): 50; Andrew Woolum, Trevor Foulk, and Amir Erez, "A Review of the Short-Term Implications of Discrete, Episodic Incivility," *Social and Personality Psychology Compass* 18, no. 1 (2024): e12918.

134 **witnessing even one small act of disrespect:** Olga Stavrova, Daniel Ehlebracht, and Kathleen D. Vohs, "Victims, Perpetrators, or Both? The Vicious Cycle of Disrespect and Cynical Beliefs About Human Nature," *Journal of Experimental Psychology: General* 149, no. 9 (2020): 1736.

136 **the importance of "active listening":** Carl R. Rogers and Richard Evans Farson, *Active Listening* (Connecticut: Martino Fine Books, 2015).

136 **"good listeners" accrue all kinds:** Avraham N. Kluger et al., "A Meta-Analytic Systematic Review and Theory of the Effects of Perceived Listening on Work Outcomes," *Journal of Business and Psychology* 39, no. 2 (2024): 295–344.

136 **actual listening is through *verbal* cues:** Hanne K. Collins et al., "Conveying and Detecting Listening During Live Conversation," *Journal of Experimental Psychology: General* (2023); Hanne K. Collins, "When Listening Is Spoken," *Current Opinion in Psychology* 47 (2022): 101402.

137 **examples of *responsiveness*:** Harry T. Reis and Shelly L. Gable, "Responsiveness," *Current Opinion in Psychology* 1 (2015): 67–71.

137 **Our minds were built to *wander*:** Malia F. Mason et al., "Wandering Minds: The Default Network and Stimulus-Independent Thought," *Science* 315, no. 5810 (2007): 393–95.

137 **42 percent of participants did not notice:** Bruno Galantucci, Simon Garrod, and Gareth Roberts, "Experimental Semiotics," *Language and Linguistics Compass* 6, no. 8 (2012): 477–93.

138 **"Colorless green ideas sleep furiously":** Gareth Roberts, Benjamin Langstein, and Bruno Galantucci, "(In) Sensitivity to Incoherence in Human Communication," *Language and Communication* 47 (2016): 15–22.

138 **listening attentively 76 percent:** Collins et al., "Conveying and Detecting Listening."

139 **the collaborative process of *grounding*:** Herbert H. Clark and Susan E. Brennan, "Grounding in Communication," in L. B. Resnick, J. M. Levine, and S. D. Teasley, eds., *Perspectives on Socially Shared Cognition*, 127–49 (New York: American Psychological Association, 1991).

139 **grounding is a three-step process:** Janet B. Bavelas et al., "The Theoretical and Research Basis of Co-Constructing Meaning in Dialogue," *Journal of Solution Focused Practices* 1, no. 2 (2014): 3.

140 **Back-channel feedback helps speakers:** Janet B. Bavelas, Linda Coates, and Trudy Johnson, "Listeners as Co-narrators," *Journal of Personality and Social Psychology* 79, no. 6 (2000): 941.

142 **Just repeat or reformulate:** Michael Yeomans et al., "Conversational Receptiveness: Improving Engagement with Opposing Views," *Organizational Behavior and Human Decision Processes* 160 (2020): 131–48.

143 **Verbal expressions of listening:** Collins, "When Listening Is Spoken."

144 **"this little glittery cloud":** Wesley Morris, "Taylor Swift: Miss Americana' Review: A Star, Surprisingly Alone," *New York Times,* January 30, 2020.

144 **Each pair got their own songwriting room:** Bitterly, Wood Brooks, Schweitzer, and Aaker, "Gender and Laughter" (working).

Chapter 6: Many Minds

152 **groups are a nightmare:** Gus Cooney et al., "The Many Minds Problem: Disclosure in Dyadic Versus Group Conversation," *Current Opinion in Psychology* 31 (2020): 22–27.

153 **In a smooth conversation:** Emma M. Templeton et al., "Fast Response Times Signal Social Connection in Conversation," *Proceedings of the National Academy of Sciences* 119, no. 4 (2022): e2116915119.

153 **Dyadic turn-taking:** Elizabeth Stokoe et al., "When Delayed Responses Are Productive: Being Persuaded Following Resistance in Conversation," *Journal of Pragmatics* 155 (2020): 70–82.

153 **subtler cues like eye contact:** So-Hyeon Shim et al., "The Impact of Leader Eye Gaze on Disparity in Member Influence: Implications for Process and Performance in Diverse Groups," *Academy of Management Journal* 64, no. 6 (2021): 1873–900; Sophie Wohltjen and Thalia Wheatley, "Eye Contact Marks the Rise and Fall of Shared Attention in Conversation," *Proceedings of the National Academy of Sciences* 118, no. 37 (2021): e2106645118.

153 **groups have much more cross-talk:** Lynn Smith-Lovin and Charles Brody, "Interruptions in Group Discussions: The Effects of Gender and Group Composition," *American Sociological Review* (1989): 424–35.

153 **as groups become larger, fewer people:** Cooney et al., "Many Minds Problem."

155 **status as the respect and prestige:** J. Stuart Bunderson and Ray E. Reagans, "Power, Status, and Learning in Organizations," *Organization Science* 22, no. 5 (2011): 1182–94.

156 **dictating who says what when:** Joey T. Cheng et al., "Two Ways to the Top: Evidence That Dominance and Prestige Are Distinct Yet Viable Avenues to Social Rank and Influence," *Journal of Personality and Social Psychology* 104, no. 1 (2013): 103; Dana R. Carney, "The Nonverbal Expression of Power, Status, and Dominance," *Current Opinion in Psychology* 33 (2020): 256–64.

157 **easier for high-status individuals to dominate:** Elad N. Sherf et al., "Centralization of Member Voice in Teams: Its Effects on Expertise Utilization and Team Performance," *Journal of Applied Psychology* 103, no. 8 (2018): 813.

158 **on the floor of the U.S. Senate:** Victoria L. Brescoll, "Who Takes the Floor and Why: Gender, Power, and Volubility in Organizations," *Administrative Science Quarterly* 56, no. 4 (2011): 622–41.

159 **women put themselves lower in rank:** Katherine Coffman et al., "Gender Stereotypes in Deliberation and Team Decisions," *Games and Economic Behavior* 129 (2021): 329–49.

160 **To voice their ideas freely:** Amy C. Edmondson and Zhike Lei, "Psychological Safety: The History, Renaissance, and Future of an Interpersonal Construct," *Annual Review of Organizational Psychology and Organizational Behavior* 1, no. 1 (2014): 23–43; Amy C. Edmondson, *The Fearless Organization: Creating Psychological Safety in the Workplace for Learning, Innovation, and Growth* (Hoboken, NJ: John Wiley & Sons, 2018).

162 **high status inhibits perspective-taking:** Adam D. Galinsky, Derek D. Rucker, and Joe C. Magee, "Power and Perspective-Taking: A Critical Examination," *Journal of Experimental Social Psychology* 67 (2016): 91–92.

162 **experiencing different levels of status:** Catarina R. Fernandes et al., "What Is Your Status Portfolio? Higher Status Variance Across Groups Increases Interpersonal Helping but Decreases Intrapersonal Well-Being," *Organizational Behavior and Human Decision Processes* 165 (2021): 56–75.

163 **openly discussing criticisms you've received:** Constantinos G. V. Coutifaris and Adam M. Grant, "Taking Your Team Behind the Curtain: The Effects of Leader Feedback-Sharing and Feedback-Seeking on Team Psychological Safety," *Organization Science* 33, no. 4 (2022): 1574–98.

163 **when high-status group members reveal:** Alison Wood Brooks et al., "Mitigating Malicious Envy: Why Successful Individuals Should Reveal Their Failures," *Journal of Experimental Psychology: General* 148, no. 4 (2019): 667.

164 **eye gaze, especially from a group leader:** Nicole Abi-Esber, Ethan Burris, and Alison Wood Brooks, "Eye Gaze" (working).

164 **Equitable eye gaze from a leader:** So-Hyeon Shim et al., "The Impact of Leader Eye Gaze on Disparity in Member Influence: Implications for Process and Performance in Diverse Groups," *Academy of Management Journal* 64, no. 6 (2021): 1873–900.

164 **resisting the urge to mostly look:** Tom Foulsham et al., "Gaze Allocation in a Dynamic Situation: Effects of Social Status and Speaking," *Cognition* 117, no. 3 (2010): 319–31.

166 **"status determines your range":** Adam Galinsky, "How to Speak up for Yourself," TED Talk, November 23, 2016.

167 **"impression management" behaviors:** Mark R. Leary and Robin M. Kowalski, "Impression Management: A Literature Review and Two-Component Model," *Psychological Bulletin* 107, no. 1 (1990): 34.

167 **likely to use technical jargon:** Zachariah C. Brown, Eric M. Anicich, and Adam D. Galinsky, "Compensatory Conspicuous Communication: Low Status Increases Jargon Use," *Organizational Behavior and Human Decision Processes* 161 (2020): 274–90.

167 **laugh politely, to appear competent:** Christopher Oveis et al., "Laughter Conveys Status," *Journal of Experimental Social Psychology* 65 (2016): 109–15.

168 **paraphrasing—repeating, summarizing:** Maria Seehausen et al., "Effects of Empathic Paraphrasing—Extrinsic Emotion Regulation in Social Conflict," *Frontiers in Psychology* 3 (2012): 31892.

168 **Lower-status group members who wield humor:** Bradford T. Bitterly, Alison Wood Brooks, and Maurice E. Schweitzer, "Risky Business: When Humor Increases and Decreases Status," *Journal of Personality and Social Psychology* 112, no. 3 (2017): 431.

172 **Great Busia as a conversational steward:** J. Elise Keith, *Where the Action Is: The Meetings That Make or Break Your Organization* (Portland, OR: Second Rise, 2018).

173 **We know a lot about starlings**: Charles Siebert, "Letter of Recommendation: Starlings," *New York Times,* February 11, 2020.

174 **Partitioning is dividing a group**: Cooney et al., "Many Minds Problem."

174 **Groups are remarkably good at adjusting**: Tanya Stivers, "Is Conversation Built for Two? The Partitioning of Social Interaction," *Research on Language and Social Interaction* 54, no. 1 (2021): 1–19.

175 **doesn't have to be centralized *or* partitioned**: Keith, *Where the Action Is.*

175 **group may repartition and recentralize**: Francesca Valsesia, Joseph C. Nunes, and Andrea Ordanini, "I Am Not Talking to You: Partitioning an Audience in an Attempt to Solve The Self-Promotion Dilemma," *Organizational Behavior and Human Decision Processes* 165 (2021): 76–89.

175 **exchanges between dominant group members**: Tanya Stivers, "Is Conversation Built for Two? The Partitioning of Social Interaction," *Research on Language and Social Interaction* 54, no. 1 (2021): 1–19.

176 **a spectrum from tight to loose**: Michele L. Gelfand, *Rule Makers, Rule Breakers: Tight and Loose Cultures and the Secret Signals That Direct Our Lives* (New York: Scribner, 2019).

Chapter 7: Difficult Moments

180 **"Don't be afraid!"**: Michael Yeomans et al.,"Conversational Receptiveness: Improving Engagement with Opposing Views," *Organizational Behavior and Human Decision Processes* 160 (2020): 131–48.

181 **Recall the chart of emotions**: James A. Russell, "A Circumplex Model of Affect," *Journal of Personality and Social Psychology* 39, no. 6 (1980): 1161.

181 **If we're too afraid, angry, or distressed**: Gerben A. Van Kleef, Carsten K. W. De Dreu, and Antony S. R. Manstead, "The Interpersonal Effects of Anger and Happiness in Negotiations," *Journal of Personality and Social Psychology* 86, no. 1 (2004): 57; Alison Wood Brooks and Maurice E. Schweitzer, "Can Nervous Nelly Negotiate? How Anxiety Causes Negotiators to Make Low First Offers, Exit Early, and Earn Less Profit," *Organizational Behavior and Human Decision Processes* 115, no. 1 (2011): 43–54.

183 **These incongruities may be harmless**: Herbert H. Clark and Susan E. Brennan, "Grounding in Communication," in L. B. Resnick, J. M. Levine, and S. D. Teasley, eds., *Perspectives on Socially Shared Cognition,* 127–49 (New York: American Psychological Association, 1991); Herbert H. Clark, *Using Language* (Cambridge: Cambridge University Press, 1996).

183 **Incongruent emotions**: Joseph P. Forgas, "Feeling and Doing: Affective Influences on Interpersonal Behavior," *Psychological Inquiry* 13, no. 1 (2002): 1–28.

183 **may indicate our differing preferences**: Luiz Pessoa, "How Do Emotion and Motivation Direct Executive Control?," *Trends in Cognitive Sciences* 13, no. 4 (2009): 160–66.

184 **aspects of our conversational compasses**: Michael Yeomans, Maurice Schweitzer, and Alison Wood Brooks, "The Conversational Circumplex: Identifying, Prioritizing, and Pursuing Informational and Relational Motives in Conversation," *Current Opinion in Psychology* 44 (2022): 293–302.

186 **people find disagreement exciting**: Deepak Malhotra, "The Desire to Win: The

Effects of Competitive Arousal on Motivation and Behavior," *Organizational Behavior and Human Decision Processes* 111, no. 2 (2010): 139–46.

186 **"it takes a lot more brain real estate"**: Joy Hirsch et al., "Interpersonal Agreement and Disagreement During Face-to-Face Dialogue: An fNIRS Investigation," *Frontiers in Human Neuroscience* 14 (2021).

187 **Deeply held personal values**: Julia A. Minson and Frances S. Chen, "Receptiveness to Opposing Views: Conceptualization and Integrative Review," *Personality and Social Psychology Review* 26, no. 2 (2022): 93–111; Julia A. Minson, Frances S. Chen, and Catherine H. Tinsley, "Why Won't You Listen to Me? Measuring Receptiveness to Opposing Views," *Management Science* 66, no. 7 (2020): 3069–94.

188 **strategies for handling disagreements**: Yeomans et al., "Conversational Receptiveness."

189 **language of civil (and uncivil) disagreement**: Ibid.

190 **the public outcry was overblown**: Michael Yeomans, "Argue Better by Signalling Your Receptiveness with These Words," *Psyche* (2021).

192 **it's important to identify the goals**: Roderick I. Swabb, Robert B. Lount, Jr., Seunghoo Chung, and Jeanne M. Brett, "Setting the Stage for Negotiations: How Superordinate Goal Dialogues Promote Trust and Joint Gain in Negotiations Between Teams," *Organizational Behavior and Human Decision Processes* 167 (2021): 157–169.

194 **Sharing personal stories is humanizing**: Emily Kubin, Kurt J. Gray, and Christian von Sikorski, "Reducing Political Dehumanization by Pairing Facts with Personal Experiences," *Political Psychology* 44, no. 5 (2023): 1119–40; Emily Kubin et al., "Personal Experiences Bridge Moral and Political Divides Better Than Facts," *Proceedings of the National Academy of Sciences* 118, no. 6 (2021): e2008389118.

196 **elevating some of these goals**: Hanne K. Collins et al., "Underestimating Counterparts' Learning Goals Impairs Conflictual Conversations," *Psychological Science* 33, no. 10 (2022): 1732–52.

197 **people systematically underestimate**: Stav Atir, Kristina A. Wald, and Nicholas Epley, "Talking with Strangers Is Surprisingly Informative," *Proceedings of the National Academy of Sciences* 119, no. 34 (2022): e2206992119.

197 **dismiss them as "bad listeners"**: Zhiying Ren and Rebecca Schaumberg, "Disagreement Gets Mistaken for Bad Listening," *Psychological Science* 35, no 5 (2024).

198 **psychological process called *belief updating***: Robin M. Hogarth and Hillel J. Einhorn, "Order Effects in Belief Updating: The Belief-Adjustment Model," *Cognitive Psychology* 24, no. 1 (1992): 1–55.

202 **warier of conversation**: J. Nicole Shelton and Jennifer A. Richeson, "Interracial Interactions: A Relational Approach," *Advances in Experimental Social Psychology* 38 (2006): 121–81.

202 **desire to avoid conversation**: J. Nicole Shelton et al., "Ironic Effects of Racial Bias During Interracial Interactions," *Psychological Science* 16, no. 5 (2005): 397–402.

202 **for those who wish to resist**: Tamar Szabó Gendler, "On the Epistemic Costs of Implicit Bias," *Philosophical Studies* 156, no. 1 (2011): 33–63.

204 **"trying to take the other person's perspective"**: Tal Eyal, Mary Steffel, and Nicholas Epley, "Perspective Mistaking: Accurately Understanding the Mind of Another Requires Getting Perspective, Not Taking Perspective," *Journal of Personality and Social Psychology* 114, no. 4 (2018): 547.

204 ***egocentric projection:*** Clayton R. Critcher and David Dunning, "Egocentric Pattern

Projection: How Implicit Personality Theories Recapitulate the Geography of the Self," *Journal of Personality and Social Psychology* 97, no. 1 (2009): 1.

204 **good proxy for making social predictions**: Zidong Zhao, Haran Sened, and Diana I. Tamir, "Egocentric Projection Is a Rational Strategy for Accurate Emotion Prediction," *Journal of Experimental Social Psychology* 109 (2023): 104521.

204 **easily accessible self-knowledge**: Jeff C. Cho and Eric D. Knowles, "I'm Like You and You're Like Me: Social Projection and Self-Stereotyping Both Help Explain Self–Other Correspondence," *Journal of Personality and Social Psychology* 104, no. 3 (2013): 444.

204 **overestimate the extent to which others share**: Gary Marks and Norman Miller, "Ten Years of Research on the False-Consensus Effect: An Empirical and Theoretical Review," *Psychological Bulletin* 102, no. 1 (1987): 72.

204 **how to communicate with novices**: Susan A. J. Birch and Paul Bloom, "The Curse of Knowledge in Reasoning About False Beliefs," *Psychological Science* 18, no. 5 (2007): 382–86; Ting Zhang, Kelly B. Harrington, and Elad N. Sherf, "The Errors of Experts: When Expertise Hinders Effective Provision and Seeking of Advice and Feedback," *Current Opinion in Psychology* 43 (2022): 91–95; Ting Zhang, Dan J. Wang, and Adam D. Galinsky, "Learning Down to Train Up: Mentors Are More Effective When They Value Insights from Below," *Academy of Management Journal* 66, no. 2 (2023): 604–37.

204 **assume that others can see our inner feelings**: Thomas Gilovich, Kenneth Savitsky, and Victoria Husted Medvec, "The Illusion of Transparency: Biased Assessments of Others' Ability to Read One's Emotional States," *Journal of Personality and Social Psychology* 75, no. 2 (1998): 332.

204 **(hot-cold empathy gap)**: Loran F. Nordgren, Kasia Banas, and Geoff MacDonald, "Empathy Gaps for Social Pain: Why People Underestimate the Pain of Social Suffering," *Journal of Personality and Social Psychology* 100, no. 1 (2011): 120.

204 **their epic perspective-taking project**: Eyal, Steffel, and Epley, "Perspective Mistaking."

206 **greatest barrier to conflict resolution overall**: William Friend and Deepak Malhotra, "Psychological Barriers to Resolving Intergroup Conflict: An Extensive Review and Consolidation of the Literature," *Negotiation Journal* 35, no. 4 (2019): 407–42.

206 **conversations about race**: Kiara L. Sanchez, David A. Kalkstein, and Gregory M. Walton, "A Threatening Opportunity: The Prospect of Conversations About Race-Related Experiences Between Black and White Friends," *Journal of Personality and Social Psychology* 122, no. 5 (2022): 853.

206 **conversations with transgender canvassers**: David Broockman and Joshua Kalla, "Durably Reducing Transphobia: A Field Experiment on Door-to-Door Canvassing," *Science* 352, no. 6282 (2016): 220–24.

206 **mere contact is humanizing**: Thomas F. Pettigrew and Linda R. Tropp, "A Meta-Analytic Test of Intergroup Contact Theory," *Journal of Personality and Social Psychology* 90, no. 5 (2006): 751.

207 **These physiological symptoms are**: Alison Wood Brooks, "Get Excited: Reappraising Pre-Performance Anxiety as Excitement," *Journal of Experimental Psychology: General* 143, no. 3 (2014): 1144.

207 **emotions in that upper-left quadrant:** Jennifer S. Lerner, Deborah A. Small, and George Loewenstein, "Heart Strings and Purse Strings: Carryover Effects of Emotions on Economic Decisions," *Psychological Science* 15, no. 5 (2004): 337–41.

209 **managing negative emotions writ large:** James J. Gross, "Emotion Regulation: Conceptual and Empirical Foundations," *Handbook of Emotion Regulation* 2 (2014): 3–20; James J. Gross, "Emotion Regulation: Current Status and Future Prospects," *Psychological Inquiry* 26, no. 1 (2015): 1–26.

209 **reframe anxiety as excitement:** Alison Wood Brooks, "Get Excited: Reappraising Pre-Performance Anxiety as Excitement," *Journal of Experimental Psychology: General* 143, no. 3 (2014): 1144; Elizabeth Baily Wolf et al., "Managing Perceptions of Distress at Work: Reframing Emotion as Passion," *Organizational Behavior and Human Decision Processes* 137 (2016): 1–12.

210 **Therapists often help their clients:** Jennifer Hettema, Julie Steele, and William R. Miller, "Motivational Interviewing," *Annual Review of Clinical Psychology* 1 (2005): 91–111.

210 **strategy of situation modification:** Stefan G. Hofmann et al., "How to Handle Anxiety: The Effects of Reappraisal, Acceptance, and Suppression Strategies on Anxious Arousal," *Behaviour Research and Therapy* 47, no. 5 (2009): 389–94; Jordi Quoidbach, Moïra Mikolajczak, and James J. Gross, "Positive Interventions: An Emotion Regulation Perspective," *Psychological Bulletin* 141, no. 3 (2015): 655.

211 **labeling others' emotions verbally:** Alisa Yu, Justin M. Berg, and Julian J. Zlatev, "Emotional Acknowledgment: How Verbalizing Others' Emotions Fosters Interpersonal Trust," *Organizational Behavior and Human Decision Processes* 164 (2021): 116–35.

211 **better emotional regulation strategies:** Yael Millgram et al., "Knowledge About the Source of Emotion Predicts Emotion-Regulation Attempts, Strategies, and Perceived Emotion-Regulation Success," *Psychological Science* 34, no. 11 (2023): 1244–55.

Chapter 8: Apologies

219 **documentary-style television series *Couples Therapy*:** Sarah Bahr, "Feeling a Bit Cramped? 'Couples Therapy' May Look Familiar," *New York Times,* April 16, 2021.

221 **example of something bigger:** Matthew Fray, "The Marriage Lesson That I Learned Too Late," *Atlantic,* April 2022.

221 **micro-harms accumulate over time:** Monnica T. Williams, "Microaggressions: Clarification, Evidence, and Impact," *Perspectives on Psychological Science* 15, no. 1 (2020): 3–26.

221 **"closeness lines":** Olivia de Recat, *Drawn Together: Illustrated True Love Stories* (New York: Voracious, 2022).

223 **pivotal inflection points:** Barry R. Schlenker and Bruce W. Darby, "The Use of Apologies in Social Predicaments," *Social Psychology Quarterly* (1981): 271–78.

223 **creating a virtuous cycle defined:** Ryan Fehr and Michele J. Gelfand, "When Apologies Work: How Matching Apology Components to Victims' Self-Construals Facilitates Forgiveness," *Organizational Behavior and Human Decision Processes* 113, no. 1 (2010): 37–50.

224 **apologies are about restoring trust:** Peter H. Kim et al., "Removing the Shadow of Suspicion: The Effects of Apology Versus Denial for Repairing Competence-Versus Integrity-Based Trust Violations," *Journal of Applied Psychology* 89, no. 1 (2004): 104.

225 **It's hard to say you're sorry:** Karina Schumann, "The Psychology of Offering an Apology: Understanding the Barriers to Apologizing and How to Overcome Them," *Current Directions in Psychological Science* 27, no. 2 (2018): 74–78.

225 **most of us are reluctant to make them:** Donald L. Ferrin, et al., "Silence Speaks Volumes: The Effectiveness of Reticence in Comparison to Apology and Denial for Responding to Integrity- and Competence-Based Trust Violations," *Journal of Applied Psychology* 92, no. 4 (2007): 893.

226 **change as meaningless "cheap talk":** Maurice E. Schweitzer, John C. Hershey, and Eric T. Bradlow, "Promises and Lies: Restoring Violated Trust," *Organizational Behavior and Human Decision Processes* 101, no. 1 (2006): 1–19.

228 **done something wrong:** Alison Wood Brooks, Hengchen Dai, and Maurice E. Schweitzer, "I'm Sorry About the Rain! Superfluous Apologies Demonstrate Empathic Concern and Increase Trust," *Social Psychological and Personality Science* 5, no. 4 (2014): 467–74.

228 **pairs who apologize easily:** Karina Schumann, Emily G. Ritchie, and Amanda Forest, "The Social Consequences of Frequent Versus Infrequent Apologizing," *Personality and Social Psychology Bulletin* 49, no. 3 (2023): 331–43.

228 **apology behavior of sixty married couples:** Karina Schumann, "Does Love Mean Never Having to Say You're Sorry? Associations Between Relationship Satisfaction, Perceived Apology Sincerity, and Forgiveness," *Journal of Social and Personal Relationships* 29, no. 7 (2012): 997–1010.

229 **catastrophic blowout:** "Deepwater Horizon: Oil Spills," Damage Assessment, Remediation, and Restoration Program, National Oceanic and Atmospheric Administration, https://darrp.noaa.gov/oil-spills/deepwater-horizon; "Environmental Costs," Encyclopædia Britannica.

230 **"environmental crisis and catastrophe":** "Gulf Oil Leak," CNN, May 27, 2010, transcript.

230 **"We're sorry for the massive disruption":** "BP Chief to Gulf Residents: 'I'm Sorry,'" CNN, May 30, 2010, transcript.

231 **when you acknowledge responsibility:** Roy J. Lewicki, Beth Polin, and Robert B. Lount, Jr., "An Exploration of the Structure of Effective Apologies," *Negotiation and Conflict Management Research* 9, no. 2 (2016): 177–96.

232 **Unfortunately, I've offended:** Ahmir Thompson, "Questlove Apologizes for Offensive Japan Comments," *Okayplayer* (2017), https://www.okayplayer.com/news/questlove-apologizes-offensive-japan-instagrams.html.

233 **Requesting forgiveness is a tempting component:** Roy J. Lewicki, and Chad Brinsfield, "Trust Repair," *Annual Review of Organizational Psychology and Organizational Behavior* 4 (2017): 287–313.

234 **over three thousand parole hearings:** Grant E. Donnelly, Hanne Collins, and Alison Wood Brooks, "How Prisoner Apologies Influence Parole Decisions" (working).

234 **seven linguistic features:** Roy J. Lewicki, Beth Polin, and Robert B. Lount, Jr., "An Exploration of the Structure of Effective Apologies," *Negotiation and Conflict Management Research* 9, no. 2 (2016): 177–96.

235 **Promises to change pull:** Maurice E. Schweitzer, John C. Hershey, and Eric T.

Bradlow, "Promises and Lies: Restoring Violated Trust," *Organizational Behavior and Human Decision Processes* 101, no. 1 (2006): 1–19.

238 **We're reluctant to:** Schumann, "Psychology of Offering an Apology."

238 **faster apologies are better:** Jimin Nam et al., "Speedy Activists: How Firm Response Time to Sociopolitical Events Influences Consumer Behavior," *Journal of Consumer Psychology* 33, no. 4 (2023): 632–44; Schumann, Ritchie, and Forest, "Social Consequences of Frequent Versus Infrequent Apologizing."

240 **only 3 percent of people gave feedback:** Nicole Abi-Esber et al., "'Just Letting You Know . . .': Underestimating Others' Desire for Constructive Feedback," *Journal of Personality and Social Psychology* 123, no. 6 (2022): 1362.

240 **feeling harsh in the moment:** Emma E. Levine and Maurice E. Schweitzer, "Are Liars Ethical? On the Tension Between Benevolence and Honesty," *Journal of Experimental Social Psychology* 53 (2014): 107–17; Emma Levine and David Munguia Gomez, "'I'm Just Being Honest': When and Why Honesty Enables Help Versus Harm," *Journal of Personality and Social Psychology* 120, no. 1 (2021): 33.

241 **stating your good intentions out loud:** Emma E. Levine et al., "Who Is Trustworthy? Predicting Trustworthy Intentions and Behavior," *Journal of Personality and Social Psychology* 115, no. 3 (2018): 468; Emma E. Levine, Annabelle R. Roberts, and Taya R. Cohen, "Difficult Conversations: Navigating the Tension Between Honesty and Benevolence," *Current Opinion in Psychology* 31 (2020): 38–43.

241 **positive feedback to come first:** Alison Wood Brooks and Leslie John, "Start with Positive Feedback" (working).

242 **focusing on forward-looking advice:** Hayley Blunden et al., "Eliciting Advice Instead of Feedback Improves Developmental Input" (working).

243 **is difficult and moving:** Lara B. Aknin and Gillian M. Sandstrom, "People Are Surprisingly Hesitant to Reach Out to Old Friends," *Communications Psychology* 2, no. 1 (2024): 34.

Epilogue

252 **third pillar of well-being:** Shigehiro Oishi and Erin C. Westgate, "A Psychologically Rich Life: Beyond Happiness and Meaning," *Psychological Review* 129, no. 4 (2022): 790.

252 **happy and meaningful life can also be *boring*:** Erin C. Westgate and Timothy D. Wilson, "Boring Thoughts and Bored Minds: The MAC Model of Boredom and Cognitive Engagement," *Psychological Review* 125, no. 5 (2018): 689.

252 **the ups and downs of life:** Jordi Quoidbach et al., "Emodiversity and the Emotional Ecosystem," *Journal of Experimental Psychology: General,* 143, no. 6 (2014): 2057; Jordan Etkin, "Choosing Variety for Joint Consumption," *Journal of Marketing Research* 53, no. 6 (2016): 1019–33.

252 **talking to a mix of people:** Hanne K. Collins et al., "Relational Diversity in Social Portfolios Predicts Well-Being," *Proceedings of the National Academy of Sciences* 119, no. 43 (2022): e2120668119.

253 **this diversity of feelings:** Jordi Quoidbach et al., "Emodiversity and the Emotional Ecosystem," *Journal of Experimental Psychology: General* 143, no. 6 (2014): 2057.

253 **you're already better at talking:** Christopher Welker et al., "Pessimistic Assessments of Ability in Informal Conversation," *Journal of Applied Social Psychology* 53,

no. 7 (2023): 555–69; Erica J. Boothby et al., "The Liking Gap in Conversations: Do People Like Us More Than We Think?," *Psychological Science* 29, no. 11 (2018): 1742–56; Adam M. Mastroianni et al., "The Liking Gap in Groups and Teams," *Organizational Behavior and Human Decision Processes* 162 (2021): 109–22; Gus Cooney, Erica J. Boothby, and Mariana Lee, "The Thought Gap After Conversation: Underestimating the Frequency of Others' Thoughts About Us," *Journal of Experimental Psychology: General* 151, no. 5 (2022): 1069.

254 **"Now that you don't have to be perfect":** John Steinbeck, *East of Eden* (Penguin, 2002).

254 **Paul Grice intended for his maxims:** Paul Grice, *Studies in the Way of Words* (Cambridge, MA: Harvard University Press, 1991).

254 **forget, overlook, or abandon our intentions:** Daniel C. Richardson, Rick Dale, and Natasha Z. Kirkham, "The Art of Conversation Is Coordination," *Psychological Science* 18, no. 5 (2007): 407–13; Adam M. Mastroianni et al., "Do Conversations End When People Want Them To?," *Proceedings of the National Academy of Sciences* 118, no. 10 (2021): e2011809118; Sophie Wohltjen and Thalia Wheatley, "Eye Contact Marks the Rise and Fall of Shared Attention in Conversation," *Proceedings of the National Academy of Sciences* 118, no. 37 (2021): e2106645118.

255 **effortful things that can become less-effortful:** Adrian M. Haith and John W. Krakauer, "The Multiple Effects of Practice: Skill, Habit and Reduced Cognitive Load," *Current Opinion in Behavioral Sciences* 20 (2018): 196–201.

255 **conversation "is the spark":** Erving Goffman, *Interaction Ritual: Essays in Face-to-Face Behavior* (Garden City, N.Y.: Doubleday, 1967).

Appendix

264 **"Jazz is a conversation":** Katie Koch, "Jazz as Conversation," *Harvard Gazette*, April 18, 2013.

264 **"You've got to learn your instrument":** Jason Pugatch, *Acting Is a Job: Real-life Lessons About the Acting Business* (New York: Allworth, 2006), 73.

264 **"A lack of seriousness":** J. Rentilly, "Kurt Vonnegut: The Last Interview with One of America's Great Men of Letters," *US Airways Magazine*, June 2007.

264 **"Play is often talked about":** Fred Rogers, *You Are Special: Neighborly Words of Wisdom from Mister Rogers* (New York: Penguin Books, 1995), 47.

264 **"We were each other's ideal audience":** Christopher Isherwood, *Lions and Shadows: An Education in the Twenties* (London: Hogarth Press, 1938), 65.

265 **"A high school student wrote":** Fred Rogers, *The World According to Mister Rogers* (New York: Hyperion, 2003), 160.

265 **"When true silence falls":** Harold Pinter, *Various Voices: Prose, Poetry, Politics 1948-1998* (London: Faber, 1998), 34.

265 **"Every difficult conversation":** *Veep*, season 5, episode 1, "Morning After," directed by Chris Addison, written by David Mandel and Armando Iannucci, aired on April 24, 2016, on HBO.

265 **"36 questions that lead to love":** Daniel Jones, "The 36 Questions That Lead to Love," *New York Times*, January 9, 2015.

INDEX

A

Aaker, Jennifer, 117, 144
Abi-Esber, Nicole, 164
acknowledgment, 191
active listening, 136
affirmation, 191–92
agreement, flagging points of, 192–93
Amabile, Teresa, 96
amygdala hijacking, 133
Aniston, Jennifer, 89
apologies
    as apex conversational skill, 225
    bad, 220, 225, 229–31
    elements of, 234–35
    examples of, 229–33, 237–38
    exercise, 263
    focus of, 230–31
    forgiveness and, 225, 228–29, 233
    power of, xxiii, 132, 223–24, 228–29, 247
    promises to change and, 235–38
    relationships and, 228–29
    reluctance to make, 225–27, 238
    research on, 234
    TALK maxims and, 247–48
    timing of, 238–40
    trust and, 224, 226, 227, 228, 236
Aron, Arthur, 265
asking
    animals and, 61–62
    bad patterns of, 81–83
    boomer-, 81–82
    cultural nuance and, 78
    excessive, 77–80
    fear of, 67–68
    lack of, 63–64
    power of, 61–62, 64–65, 69–70, 149
    relationships and, 62–63
    See also questions
Atir, Stav, 197
Austin, J. L., 11, 16

B

back-channel feedback, 138–41
Bagdonas, Naomi, 105
Bailey, David, 211
Bavelas, Janet, 139, 140
belief updating, 198
Benign Violation Theory of Humor, 101–2
Benjamin, Walter, 96
Berg, Justin, 211
Bitterly, Brad, 100, 144
Bohns, Vanessa, 115
boomerasking, 81–82
Boothby, Erica, 115, 253
Boston Celtics, xxi
BP, 229–31
Brady, Tom, 199–201
Brescoll, Victoria, 158
"broaden and build" theory, 99
Broockman, David, 206
Burris, Ethan, 164
Bush, George W., 89

C

call-backs, 110–11, 141
candidate answers, 85
care, worthy of, 132–33
Carell, Steve, 119
Carnegie, Dale, 10
Chan, Elaine, 114
Chat Circle exercise, 24–27, 45

Clinton, Bill, 76
closed questions, 84–85
closeness lines, 221–23, 255
Coffman, Katie, 159
cognitive offloading, 34
cognitive reappraisal, 209
Colbert, Stephen, 118–22, 126–27, 138–39, 140, 142, 144
Collins, Hanne, 138, 189, 234
companionable silence, 263
compare-and-contrast strategy, 105–6
compliments, 113–15, 262
context
    changes in, xiii, 16
    humor and, 99–100
    importance of, xiii
    shifting, 109, 210
conversational compass, 12–16
conversational receptiveness. *See* receptiveness
conversational stewardship, 172–77
Conversation Analysis, 19
conversations
    co-creation and, 31
    complexity of, xiii, 8–9, 21
    context and, xiii
    cooperative principle and, 17, 258
    as coordination game, 6–10, 258
    dating and, 36–38, 88–93
    definitions of, 1–2
    direct communication vs., 10–11
    diversity and, 252–53
    egocentrism and, 124
    ending, 9
    examples of, 1
    feedback on, xv–xviii
    gender differences in, 20
    goals for, 257, 259–60
    Grice's maxims of, 16–18, 254, 258
    history of, 2–6
    increasing effectiveness of, xiv–xv, 21–22, 254
    as jazz of human exchange, 9–10
    Kant's rules for, 3–5
    keeping alive, 86–87
    masters of, 264
    meaningful, 259
    micro-decisions in, 9
    missteps in, xi–xiii
    mutual pleasure and, 2–3
    naturalness of, 11
    nervousness and, 253
    personal history of, 257
    practicing, 253, 255, 259–64
    prepping for, 24–27, 32–36, 46–47, 260
    purposes for, xiv, 11–16
    relationships and, xix–xx, 255
    research on, xviii–xix, 18–21, 88–89
    self-assessments of skill at, 253–54
    shared understanding and, 125–26
    synchronous, 1
    turn-taking in, 153
    See also apologies; asking; difficult conversations; group conversations; kindness; levity; topics
Cooney, Gus, 9
Cooper, Anderson, 118–22, 126, 138–39, 140, 142, 144
Cooper, Carter, 118
Cooper, Wyatt, 118
coordination games
    conversation as, 6–10
    cooperative vs. noncooperative, 6, 15
    definition of, 6
    examples of, 6–7
Coutifaris, Constantinos, 163

D

Dai, Hengchen, 228
dating conversations, 36–38, 88–93, 108–9
David, Larry, 105
Deepwater Horizon oil spill, 229–31
de-escalation, 261
de Recat, Olivia, 221–22, 255
de Staël, Germaine, 2
Dickens, Charles, 5
differences, levels of, 182–86
difficult conversations
    challenge of, 186
    differences and, 182–86, 199–203
    emotions and, xviii, 180–83, 207–11

example of, 211–17
exercises, 179–80, 258
other orientation and, 205–7
overheated, 207–11
perspective-getting and, 203–5, 206
TALK maxims and, 187–88
See also receptiveness
digital communication audit, 259
digital identity, 259
disagreements. *See* difficult conversations
disclosure, 44–45, 61, 97, 98
disengagement vs. engagement, 94–95
disrespect
  contagion of, 133–35
  language and, 127, 129, 130, 131
Di Stasi, Matteo, 142
Donnelly, Grant, 234
duologues, 137

E

eavesdropping, 18–21
Edmondson, Amy, 160
egocentric projection, 204
egocentrism, 123–24
emotional tone, 131–32
emotions
  difficult conversations and, xviii, 180–83, 207–11
  dimensions of, 94–95
  managing negative, 209–11
  power of positive, 96–97, 109
  understanding source of, 211
engagement vs. disengagement, 94–95
Epley, Nick, 203–4
escalation, 261
explanation words, avoiding, 194
Eyal, Tal, 203–4
eye gaze, power of, 164, 263

F

feedback
  back-channel, 138–41
  challenge, 263
  effectiveness of, xv, xvi–xviii
  giving, 241–42
  importance of, 240
  lack of, xv–xvi, 240
  receiving, 241, 242
  sandwich, 241–42
  timing of, 241
Fernandes, Catarina, 162
first dates, 88, 93
Fisher, Carrie, 56–61, 62, 65
Fisher, Eddie, 56
flattery, 113–15
focal points, 7
follow-up questions, 72, 74–77, 85–86, 141, 261
Ford, Harrison, 58–60, 62
forgiveness, 225, 228–29, 233, 234
Foulk, Trevor, 133
Franklin, Aretha, 127
Frederickson, Barbara, 99
Frei, Frances, 242

G

Galantucci, Bruno, 137
Galinsky, Adam, 162, 166
gender differences, 20, 158–60
Gendler, Tamar Szabó, 202
Gilovich, Tom, 253
Goethe, Johann Wolfgang von, 2
Goffman, Erving, 18–19, 20, 100, 167, 255
Goldenberg, Amit, 211
Golden Rule, 203, 205
Gong, 79–80
gotcha questions, 83
Grand Central Terminal, 7
Grande, Ariana, xiii
Grant, Adam, 163
Greenwald, Rachel, 63, 64, 95, 116
Grice, Paul, xxii, 16–18, 21, 254, 258
Gross, Terry, 57–61, 62, 65, 75, 76, 77
grounding, 139
group conversations
  airtime sharing in, 153, 157–58
  centralizing, 174–75
  challenges of, 153–54, 177
  conflict and, 152–53
  examples of, 152, 154

group conversations (*cont.*)
- fostering inclusion in, 160–69
- gender dynamics and, 158–60
- partitioning, 174, 175
- party panic and, 150–51
- positive aspects of, 151, 152
- psychological safety and, 160
- status hierarchies and, 155–69
- structuring, 172–77
- tight vs. loose, 176–77
- turn-taking in, 153

Guralnik, Orna, 219, 220, 223, 226, 237, 238

H

Hayward, Tony, 230–31
hedges, 129, 193
Hirsch, Joy, 186
Hochschild, Arlie, 9
Houston, Whitney, 251
Hume, David, 2
humor
- affiliative vs. aggressive, 102–4
- Benign Violation Theory of, 101–2
- context-sensitive nature of, 99–100
- effectiveness of, 100–101
- styles of, 102, 106–7

See also laughter; levity

I

identities
- differences in, 184–85, 199–207
- digital, 259
- mapping, 258–59
- plurality of, 130–31

impression management, 167
interrupting, 142
introductory questions, 71, 72
Isherwood, Christopher, 264–65

J

John, Leslie, 97, 241
Jolie, Angelina, 89
Jurafsky, Dan, 89

K

Kalla, Joshua, 206
Kant, Immanuel, 3–5, 9, 10, 18, 21, 173
Kaplan, Abraham, 137
Keith, Elise, 172, 175
Keller, Ric, 102–3
Keysar, Boaz, 124–26, 135
kindness
- example of, 144–48
- importance of, 122–23, 149
- listening and, 135–37
- practicing, 123
- respectful language and, 127–33
- trust and, 224–25

Kristal, Ariella, 138, 242

L

Lady Gaga, 144
language
- disrespectful, 127, 129, 130, 131, 132
- para-, 104
- positive and negative, 131–32
- respectful, 127–33

laughter
- power of, 112–13
- purposes of, 108–9, 113
- research on, 107–8, 113
- triggers for, 108

Levi, Primo, 28, 53, 189
levity
- age and, 98
- attempts at, 107, 109
- call-backs and, 110–11
- example of, 89–93
- importance of, 149
- mindset and, 104–6
- power of, 93–97, 107, 118–19
- reframing with, 261–62
- supporting, 112–13

See also humor; laughter

Lewicki, Roy, 231, 234
Lewinsky, Monica, 76–77, 130
listening
- active, 136
- back-channel feedback and, 138–41
- exercise, 262

lack of, 137–38
verbal cues of, 136–37, 143–44
See also responsive listening

M

Marsalis, Wynton, 10, 264
Martin, Sean, xix
Mastroianni, Adam, 9, 192
McCarthy, Timothy, 111
McFarland, Daniel, 89
McGraw, Peter, 101, 104
Mehl, Matthias, 20
micro-aggressions, 127
micro-decisions, 9
micro-kindnesses, 127
Millgram, Yael, 211
Minson, Julia, 138, 189
mirror questions, 71–73
mood elevator, 95
Mulaney, John, 105–6

N

Nahum, Mordu, 28
names, use of, 130
natural language processing (NLP), 20
Naturalness, Myth of, 32, 34, 86, 254
negative framing, 193
Never-ending Follow-ups exercise, 75, 261
news, as conversation topic, 40
nicknames, 130
Nock, Matthew, 211

O

Oishi, Shigehiro, 252
open-ended questions, 84–85
Oswald, Andrew, 96
other orientation, 205–7

P

paralanguage, 104
paraphrasing, 262
Parker, Charlie, 264
Parker, James, 42

parole process, 233–36
party panic, 150–51
perspective-getting, 203–5, 206
Piaget, Jean, 123
Pinter, Harold, 265
Pitt, Brad, 89
Pomerantz, Anita, 85
positive framing, 193–94
Post, Emily, 10
pregnancy, asking about, 67–68
Prisoner's Dilemma, 15, 184
Provine, Robert, 107–9, 113
psychological richness, 252
psychological safety, 97, 102, 160, 224–25

Q

questions
asking too many, 77–80
bad, 81–83
closed vs. open-ended, 84–85
lists of, 265–70
motives for, 70–71, 81–83
repeated, 83
sensitive, 65–69, 82, 260–61
types of, 71–77
See also asking
Questlove, 231–33, 234
quotes, favorite, 264–65

R

Ranganath, Rajesh, 89
receptiveness
mindset for, 196–98
recipe for, 189–96
Redding, Otis, 127
reflection exercises, 257–59
reframing, 209–10, 261–62
Relationship Reboot exercise, 243–47, 263
relationships
apologies and, 228–29
closeness lines and, 221–23, 255
conversations and, xix–xx, 255
feedback and, 240
importance of, xix

relationships *(cont.)*
  questions and, 62–63
  repairing, 242–47
repeated questions, 83
respect, 127–33. *See also* disrespect
responsive listening
  back-channel feedback and, 138–41
  benefits of, 143–44
  callbacks and, 141
  follow-up questions and, 141
  importance of, 137
  interrupting and, 142
  repeating and, 142
Reynolds, Debbie, 56
Richeson, Jennifer, 202
richness, psychological, 252
Rickles, Don, 103
Roberts, Gareth, 137
Rogers, Fred, 264, 265

S

safety, psychological, 97, 102, 160, 224–25
Sanchez, Kiara, 206
Sandstrom, Gillian, 20
Schelling, Thomas, 7, 8, 15
Schumann, Karina, 228–29
Schweitzer, Maurice, 100, 144, 228
Seinfeld, Jerry, 104, 105
Seinfeld Effect, 104–5
self-deprecation, 103
Sen, Amartya, 130
Sengupta, Jaideep, 114
Shaw, George Bernard, 125
Shelton, Nicole, 202
silence
  retreat, 263
  shared, 263
Silverman, Sarah, 105
Sinatra, Frank, 103
situation modification, 109, 210
small talk
  doom loop of, 37–38, 41
  power of, 36–41
  as steppingstone, 42–46
Smith, Adam, 2, 5, 6, 11

Spender, Stephen, 28–29, 30, 52
status hierarchies, 155–69
Stefani, Gwen, 89
Steffel, Mary, 203–4
Steinbeck, John, 254
Stewart, Jon, 119
stories
  developing stable of, 111
  sharing, 194
Swaab, Roderick, 192
Swift, Jonathan, 2
Swift, Taylor, 144
system 1 vs. system 2 thinking, 34, 39

T

TALK maxims
  apologies and, 247–48
  difficult conversations and, 187–88
  intuitiveness of, 21
  origin of, 21
  overview of, xxii
  practicing, 259–64
  as reminders, 147, 254–55
  situations that test, 149
  teaching, 264
  trust and, 224
  See also asking; kindness; levity; topics
Taylor, Elizabeth, 56, 103
This American Life, 40–41
Tocqueville, Alexis de, 5
topics
  abstract vs. concrete, 44
  co-creation and, 31
  commonalities as, 45
  definition of, 29
  disclosure and, 44–45
  disinterest and, 50–52
  "good" vs. "bad," 30–31, 53–55
  importance of, 29, 149
  leading and following, 167–68, 260
  lists of, 265–70
  management of, 31–32, 52–53
  prepping, 24–27, 32–36, 46–47, 251, 260

pyramid of, 43–46
stagnant, 47–48
switching, 48–53, 72, 73–74, 75–76
universal, 39–40
weird, 111–12
Tripp, Linda, 76
trust
apologies and, 224, 226, 227, 228, 236
establishing, 97, 224–25
restoring, 224, 242–47
TALK maxims and, 224

V
validation, 262
Vanderbilt, Gloria, 118
variety, need for, 252
vipassana, 263
Vohs, Kathleen, 134
Vonnegut, Kurt, 264

W
Walker, Jesse, 253
Walters, Barbara, 76–77, 130
Warren, Caleb, 101
weather, as conversation topic, 37–38
weirdness, 111–12
Welker, Christopher, 253
well-being, pillars of, 252
Westgate, Erin, 252
Wigilia, 169–74, 177
Winfrey, Oprah, 76, 77, 261
Woolf, Virginia, 28–29, 30, 52

Y
Yeomans, Mike, 30, 74, 189, 242, 267

Z
Zlatev, Julian, 211
ZQs (Zero Questions), 63–64, 86

On a station platform, with nothing to read,
and a four-hour train journey stretching ahead of him...

That's where the story began for Penguin founder Allen Lane.
With only 'shabby reprints of shoddy novels' on offer,
he resolved to make better books for readers everywhere.

By the time his train pulled into London, the idea was formed.
He would bring the best writing, in stylish and affordable
formats, to everyone. His books would be sold in bookstores,
stationers and tobacconists, for no more than the price
of a ten-pack of cigarettes.

And on every book would be a Penguin, a bird with a certain
'dignified flippancy', and a friendly invitation to anyone who
wished to spend their time reading.

In 1935, the first ten Penguin paperbacks were published.
Just a year later, three million Penguins had made their
way onto our shelves.

Reading was changed forever.

—

A lot has changed since 1935, including Penguin, but in the
most important ways we're still the same. We still believe that
books and reading are for everyone. And we still believe that
whether you're seeking an afternoon's escape, a vigorous debate
or a soothing bedtime story, all possibilities open with a book.

Whoever you are, whatever you're looking for,
you can find it with Penguin.